Lecture Notes in Earth Sciences

ctd. on inside back cover

Lecture Notes in Earth Sciences

Edited by Somdev Bhattacharji, Gerald M. Friedman,
Horst J. Neugebauer and Adolf Seilacher

34

Ruediger Stein

Accumulation of Organic Carbon in Marine Sediments

Results from the Deep Sea Drilling Project/
Ocean Drilling Program
(DSDP/ODP)

Springer-Verlag
Berlin Heidelberg GmbH

Author

Priv. Doz. Dr. Ruediger Stein
Institut für Geowissenschaften und Lithosphärenforschung
Justus-Liebig-Universität Giessen
Senckenbergstr. 3, D-6300 Giessen, F. R. Germany

Present address:
Alfred-Wegener-Institut für Polar- und Meeresforschung
Columbusstrasse, D-2850 Bremerhaven, F. R. Germany

ISBN 978-3-540-53813-4   ISBN 978-3-540-46307-8 (eBook)
DOI 10.1007/978-3-540-46307-8

2132/3140-543210 – Printed on acid-free paper

Preface

In 1984, I had had the chance to join the organic geochemistry group of Prof. Dr. D. Welte at the Institute for Petroleum and Organic Geochemistry, KFA Jülich, where my first encounter with organic geochemistry took place. There, I learned - as marine geologist - to talk to and work with geochemists and to understand the importance of multidisciplinary cooperation. For this possibility, I gratefully thank Prof. Welte as well as Dr. J. Rullkötter.

At Jülich, I started to study the organic geochemistry of DSDP Mesozoic black shales and their formation in relationship to paleoclimate, paleoceanography, and paleogeographic boundary conditions. During this work on depositional environments of black shales, the idea to study more recent (Quaternary and Neogene) organic-carbon-rich sediments from very different environments as case studies was born. Thus, in the recent past, I participated three ODP-Legs on which late Cenozoic organic-carbon-rich sediments were drilled: ODP-Leg 105 (Baffin Bay and Labrador Sea; August to October 1985), ODP-Leg 108 (subtropical-tropical Northeast Atlantic Ocean; February and March 1986), and ODP-Leg 128 (Japan Sea; August to October 1989). The results of the sedimentological and organic-geochemical investigations of these sediments allowed detailed reconstructions of mechanisms controlling organic carbon accumulation and provided important informations about the evolution of paleoclimate and paleoceanic circulation in the study areas. These results may also help to understand the formation of fossil organic-carbon-rich sedimentary rocks, i.e., petroleum source rocks.

At the KFA Jülich, large parts of the determinations of quality parameters of the organic matter were performed in cooperation with Drs. Lo ten Haven, R. Littke, P. Müller, and J. Rullkötter. For this cooperation which also continued after my change to the IGL, Giessen University, in 1986 and which exists until today, and for the numerous stimulating discussions I gratefully thank these friends and collegues. I would like to thank also Dr. P. Müller (now Bremen University) for determinations of stable carbon isotopes of organic carbon fractions and discussion of the data. Th. Wolf (GEOMAR Kiel) provided unpublished organic carbon data from Site 646, which is gratefully acknowledged. I also thank Prof. Dr. M. Sarnthein and R. Tiedemann (Kiel University) for cooperation and discussion of Leg 108 data. For technical assistance, I thank my students at IGL, Giessen University, D. Brosinsky, M. Dersch, Th. Hofmann, S. Knecht, U. Mann, and R. Stax, without their help the processing of the enormous amounts of sediment samples would not have been possible. Photographic work has been performed by G. Appel (IGL, Giessen University) which is also gratefully acknowledged.

For exchange of ideas, discussions, and technical help I thank all my shipboard collegues and marine technicians from ODP-Legs 105, 108, and 128, with special thanks to my friend, collegue, and 105-participant Dr. G. Bohrmann (AWI Bremerhaven). Stimulating discussions with numerous collegues during the Dahlem Workshop on

"Productivity of the Ocean: Past and Present" (Berlin, April 24-29, 1988) were also most helpful.

For the critical reviews of the manuscript and for numerous constructive suggestions for improvement of the manuscript I would like to thank Drs. G.M. Friedman (Northeastern Science Foundation, Troy, New York), P. Müller (Bremen University), J. Rullkötter (KFA Jülich), and J. Thiede (GEOMAR Kiel).

This study was supported by the "Deutsche Forschungsgemeinschaft" which is gratefully acknowledged.

The manuscript has been accepted as "Habilitationsschrift" by the Faculty of Geosciences and Geography of the Justus-Liebig-University in November 1990.

Ruediger Stein
Institut für Geowissenschaften und Lithosphärenforschung (IGL)
Justus-Liebig-University Giessen
November 1990.

Abstract

Starting from a more general discussion of mechanisms controlling organic carbon deposition in marine environments and indicators useful for paleoenvironmental reconstructions, this study concentrates on detailed organic-geochemical and sedimentological investigations of late Cenozoic deep-sea sediments from (1) the Baffin Bay and the Labrador Sea (ODP-Leg 105), (2) the upwelling area off Northwest Africa (ODP-Leg 108), and (3) the Sea of Japan (ODP-Leg 128). Of major interest are short- as well as long-term changes in organic carbon accumulation during the past 20 m.y.

As shown in the data from ODP-Legs 105, 108, and 128, sediments characterized by similar high organic carbon contents can be deposited in very different environments. Thus, simple total organic carbon data do not allow (i) to distinguish between different factors controlling organic carbon enrichment and (ii) to reconstruct the depositional history of these sediments. Data on both quantity <u>and</u> composition of the organic matter, however, provide important informations about the depositional environment and allow detailed reconstructions of the evolution of paleoclimate, paleoceanic circulation, and paleoproductivity in these areas. The results have significant implications for quantitative models of the mechanisms of climatic change. Furthermore, the data may also help to explain the formation of fossil black shales, i.e., hydrocarbon source rocks.

(1) Baffin Bay and Labrador Sea

The Miocene to Quaternary sediments at Baffin Bay Site 645 are characterized by relatively high organic carbon contents, most of which range from 0.5% to almost 3%. This organic carbon enrichment was mainly controlled by increased supply of terrigenous organic matter throughout the entire time interval. Two distinct maxima were identified: (i) a middle Miocene maximum, possibly reflecting a dense vegetation cover and fluvial sediment supply from adjacent islands, that decreased during late Miocene and early Pliocene time because of expansion of tundra vegetation due to global climatic deterioration; (ii) a late Pliocene-Pleistocene maximum possibly caused by glacial erosion and meltwater outwash. Significant amounts of marine organic carbon were accumulated in western Baffin Bay during middle Miocene time, indicating higher surface-water productivity (up to about 150 gC m^{-2} y^{-1}) resulted from the inflow of cold and nutrient-rich Arctic water masses. The decrease in average surface-water productivity to values similar to those of the modern Baffin Bay was recorded during the late Miocene and was probably caused by the development of a seasonal sea-ice cover.

At Labrador Sea Sites 646 and 647, organic carbon contents are low varying between 0.10% and 0.75%; the origin of most of the organic matter probably is marine. A major increase in organic carbon accumulation at Site 646 at about 7.2 Ma may indicate increased surface-water productivity triggered by the onset of the cold East-Greenland Current system. Near 2.4 Ma, i.e., parallel to the development of major Northern Hemisphere Glaciation, accumulation rates of both organic carbon and biogenic opal decreased, suggesting a reduced surface-water productivity because of the development of closed seasonal sea-ice cover in the northern Labrador Sea. The influence of varying sea-ice cover on surface-water productivity is also documented in the short-term glacial/interglacial fluctuations in organic carbon deposition at Sites 646 and 647.

(2) Upwelling areas off Northwest Africa

The upper Pliocene-Quaternary sediments at coastal-upwelling Site 658 are characterized by high organic carbon contents of 4%; the organic matter is a mixture of marine and terrigenous material with a dominance of the marine proportion. The upper Miocene to Quaternary pelagic sediments from close-by non-upwelling Sites 657 and 659, on the other hand, display low organic carbon values of less than 0.5%. Only in turbidites and slumps occasionally intercalated at the latter two sites, high organic carbon values of up to 3% occur. The high accumulation rates of marine organic carbon recorded at Site 658 reflect the high-productivity upwelling environment. Paleoproductivity varies between 100 and 400 gC m^{-2} y^{-1} during the past 3.6 m.y. and is clearly triggered by changes in global climate. However, there is no simple relationship between climate and organic carbon supply, i.e., it is not possble to postulate that productivity was generally higher at Site 658 during glacials than during interglacials or vice versa. Changes in the relative importance between upwelling activity (which was increased during glacial intervals) and fluvial nutrient supply (which was increased during interglacial intervals) may have caused the complex productivity record at Site 658. Most of the maximum productivity values, for example, were recorded at peak interglacials and at terminations indicating the importance of local fluvial nutrient supply at Site 658. Near 0.5 Ma, a long-term decrease in paleoproductivity occurs, probably indicating a decrease in fluvial nutrient supply and/or a change in nutrient content of the upwelled waters. The former explanation is supported by the contemporaneous decrease in terrigenous organic carbon and (river-borne) clay supply suggesting an increase in long-term aridity in the Central Sahara.

At Site 660, underneath the Northern Equatorial Divergence Zone, (marine) organic carbon values of up to 1.5% were recorded in upper Pliocene-Quaternary sediments. During the last 2.5 Ma, the glacial sediments are carbonate-lean and enriched in organic carbon probably caused by the influence of a carbonate-dissolving and oxygen-poor deep-water mass.

(3) Sea of Japan

Based on preliminary results of organic-geochemical investigations, the Miocene to Quaternary sediments from ODP-Sites 798 (Oki Ridge) and 799 (Kita-Yamato-Trough) are characterized by high organic carbon contents of up to 6%; the organic matter is a mixture between marine and terrigenous material. Dominant mechanisms controlling (marine) organic carbon enrichments are probably high-surface water productivity and increased preservations rates under anoxic deep-water conditions. In the lower Pliocene sediments at Site 798 and the Miocene to Quaternary sediments at Site 799, rapid burial of organic carbon in turbidites may have occurred episodically. Distinct cycles of dark laminated sediments with organic carbon values of more than 5% and light bioturbated to homogenous sediments with lower organic carbon contents indicate dramatic short-term paleoceanographic variations. More detailed records of accumulation rates of marine and terrigenous organic carbon and biogenic opal as well as a detailed oxygen isotope stratigraphy are required for a more precise reconstruction of the environmental history of the Sea of Japan through late Cenozoic time.

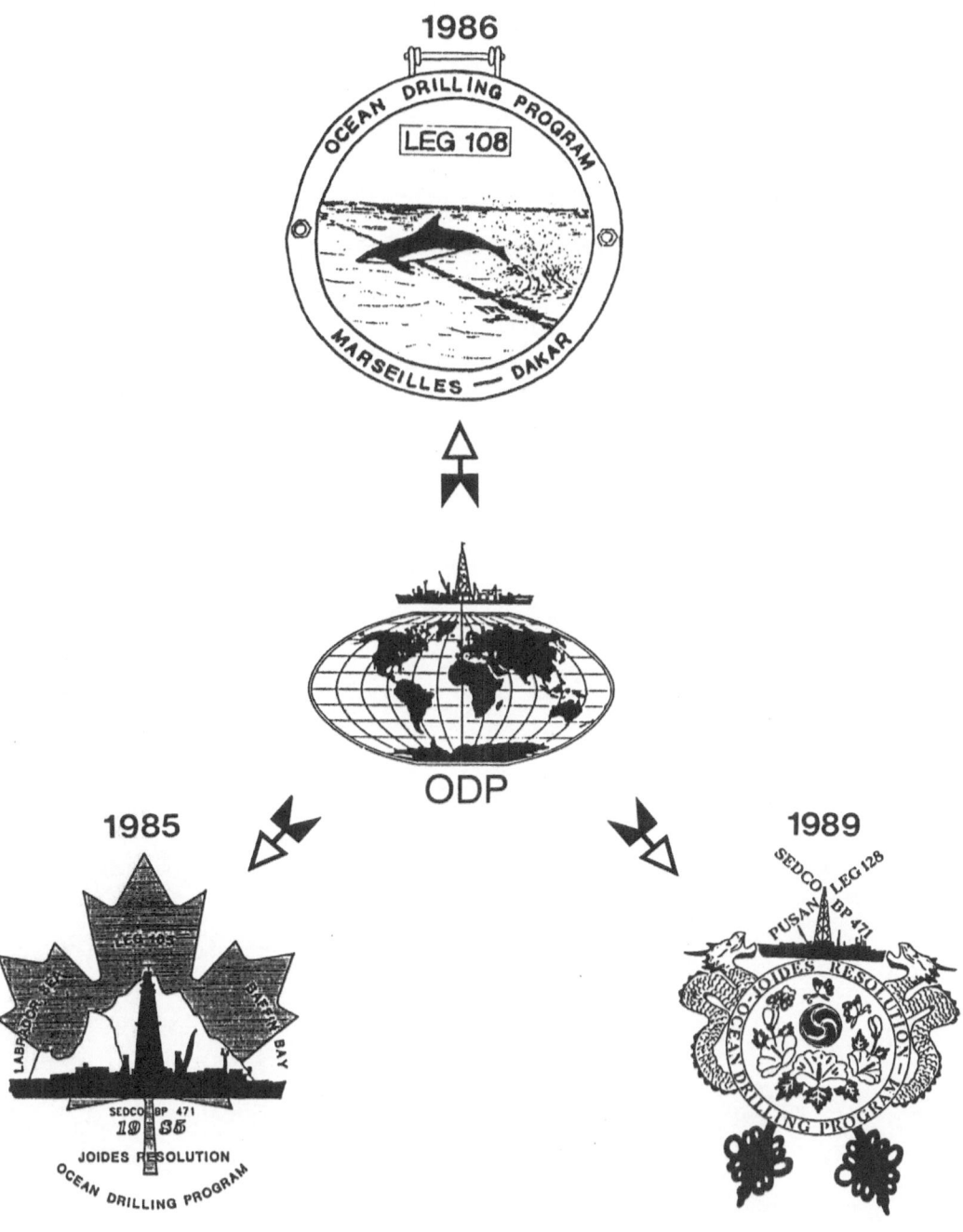

To Beate,
Annika, Jan, and Hauke

Contents

1. Introduction

The distribution of organic carbon in modern marine sediments is very complex, but also shows some regularities (Fig. 1; Premuzic et al., 1982; Romankevich, 1984). In general, shelf/upper slope sediments and sediments from marginal seas are enriched in organic carbon, whereas sediments from the central open-ocean regions display distinctly lower organic carbon contents. In detail, however, distinct differences from region to region are obvious. The situation becomes more complex if one also considers the composition of the organic matter, i.e., the proportion of terrigenous and marine organic matter deposited in the different environments. These differences in amount and composition of organic matter occur because very different mechanisms control the accumulation of organic matter in the marine realm, such as, for example, the supply from land ("allochthonous organic matter") and the production in the ocean by photosynthesizing plants ("autochthonous organic matter") (see Chapter 2). Since these mechanisms are environment-dependant, i.e., they are controlled by climatic and paleoceanographic factors, the organic carbon record of marine sediments may reflect the environment during times of deposition. Thus, changes in organic carbon contents of marine sediments through time may reflect changes in the paleoenvironment. For example, during mid-Cretaceous times organic-carbon-rich sediments were much more common and widespread and also occur in the deep open-ocean basins, suggesting a very different depositional environment than that of the modern ocean (e.g., Schlanger and Jenkyns, 1976; Tissot et al., 1979; Arthur et al., 1984; Rullkötter et al., 1984; Herbin et al., 1986; Stein et al., 1986; Sliter, 1989).

To understand the spatial distribution of organic carbon in modern sediments and its change through time, therefore, is of interest for several reasons:

(1) The study of marine organic matter and its change through time may give information about changes in surface-water productivity and/or changes in oxygen content of deep water (e.g., Müller and Suess, 1979; Emerson et al., 1985; Stein, 1986a; Sarnthein et al., 1987; Canfield, 1989; Berger et al., 1989 and reference therein). Since surface-water productivity influences the exchange of CO_2 between ocean and atmosphere, changes in bioproductivity may affect the concentration of atmospheric CO_2 which is an important factor controlling the global climate (Berger and Keir, 1984; Siegenthaler and Wenk, 1984; Genthon et al., 1987). Thus, these data have significant implications for quantitative models of the mechanisms of climatic change.

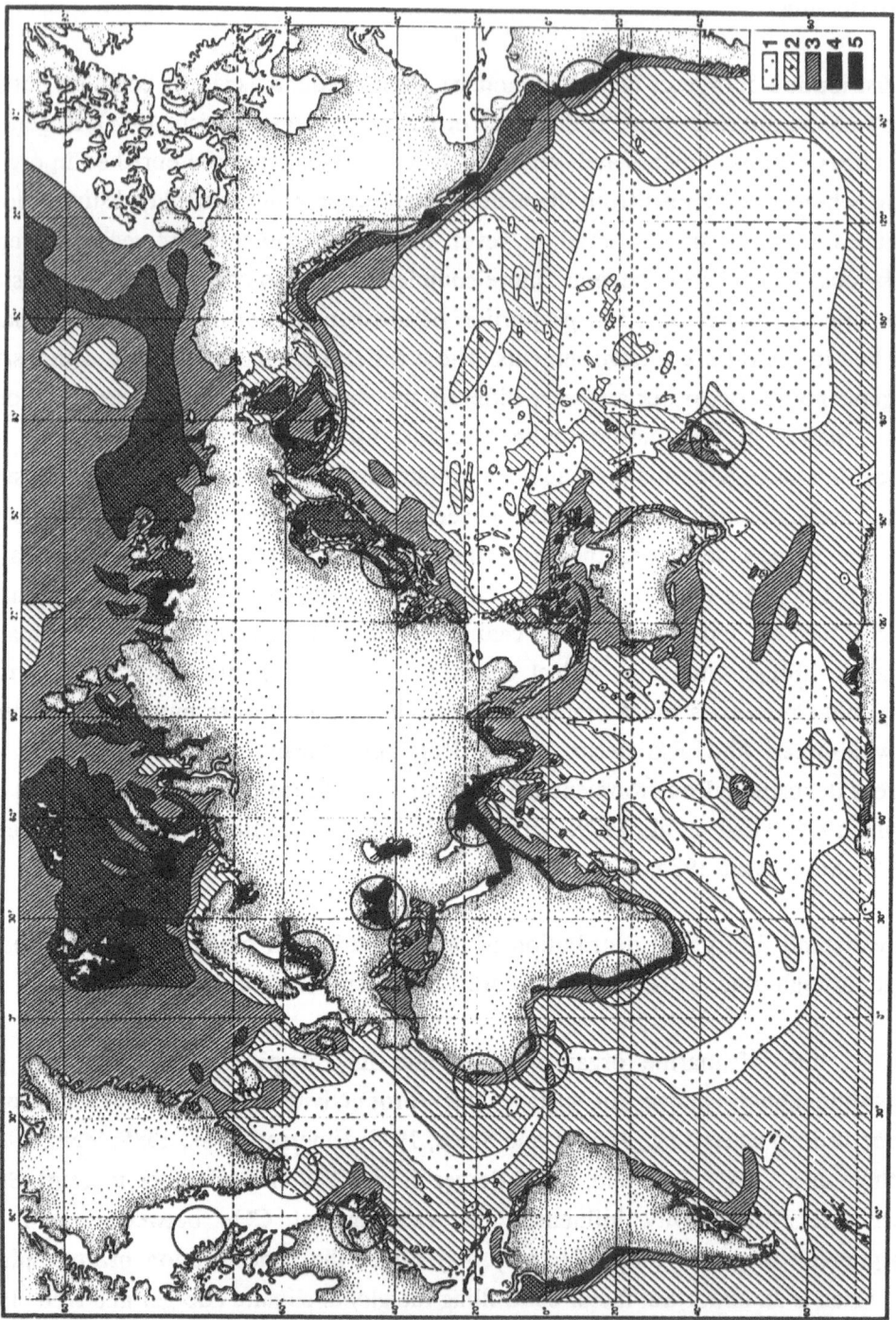

Fig. 1: Distribution of organic carbon in surface sediments of the world ocean. 1 = < 0.25, 2 = 0.25-0.5, 3 = 0.5-1, 4 = 1-2, and 5 = > 2 % (from Romankevich, 1984). Encircled areas are discussed/mentioned in the text.

(2) The study of terrigenous organic matter in marine sediments and its change through time may give information about the climatic evolution of the surrounding continents (e.g., Tissot et al., 1979; Arthur et al., 1984; Stein et al., 1986, 1989a).

(3) The investigation of the quantity and quality of organic matter in (Cenozoic) marine sediments from different environments may yield depositional models, which may help to explain the formation of fossil organic-carbon-rich sediments and sedimentary rocks (i.e., black shales). Throughout the Phanerozoic, several intervals of widespread occurrence of black shales have been recorded, such as the Frasnian/Famennian Black Shales (e.g., Buggisch, 1972; Walliser, 1986; Sandberg et al., 1988), the Permian Kupferschiefer (e.g., Paul, 1982; Kulick et al., 1984), the Jurassic Posidonia Shale (e.g., Jenkyns, 1985; Riegel et al., 1986), and the Cretaceous Black Shales (e.g., Schlanger and Jenkyns, 1976; Rullkötter et al., 1983; Arthur et al., 1984, 1988; Herbin et al., 1986; Stein et al., 1986, 1989c). Since these black shales are major petroleum source rocks, understanding their formation has not only a scientific value, but may also be of interest for petroleum source rock prospection.

This study of organic carbon accumulation in marine environments was performed on core material from the Deep Sea Drilling Project / Ocean Drilling Program (DSDP/ODP). Thus, the shallow-marine shelf areas, a further important sink for organic carbon (e.g., Walsh, 1989), are not considered here. The ODP drilling techniques provide the unique possibility to drill virtually undisturbed and complete sediment sections of Holocene/Pleistocene and pre-Pleistocene age in areas of high sedimentation rate. This allows high-resolution studies of paleoenvironmental changes through Cenozoic times, i.e., studies which are restricted to Holocene/late Pleistocene time intervals when using normal gravity or piston core techniques.

Starting from a discussion of mechanisms controlling organic carbon deposition (Chapter 2) and the significance of paleoenvironment indicators in marine sediments (Chapter 3), this study concentrates on results from ODP-Leg 105 (Baffin Bay and Labrador Sea; Chapter 5), ODP-Leg 108 (upwelling area off Northwest Africa; Chapter 6), and ODP-Leg 128 (Sea of Japan; Chapter 8) (Fig. 2). Legs 105, 108, and 128 were chosen because during these drilling campaigns several sites characterized by organic-carbon-rich sediments were drilled in very different environments (i.e., high terrigenous

4

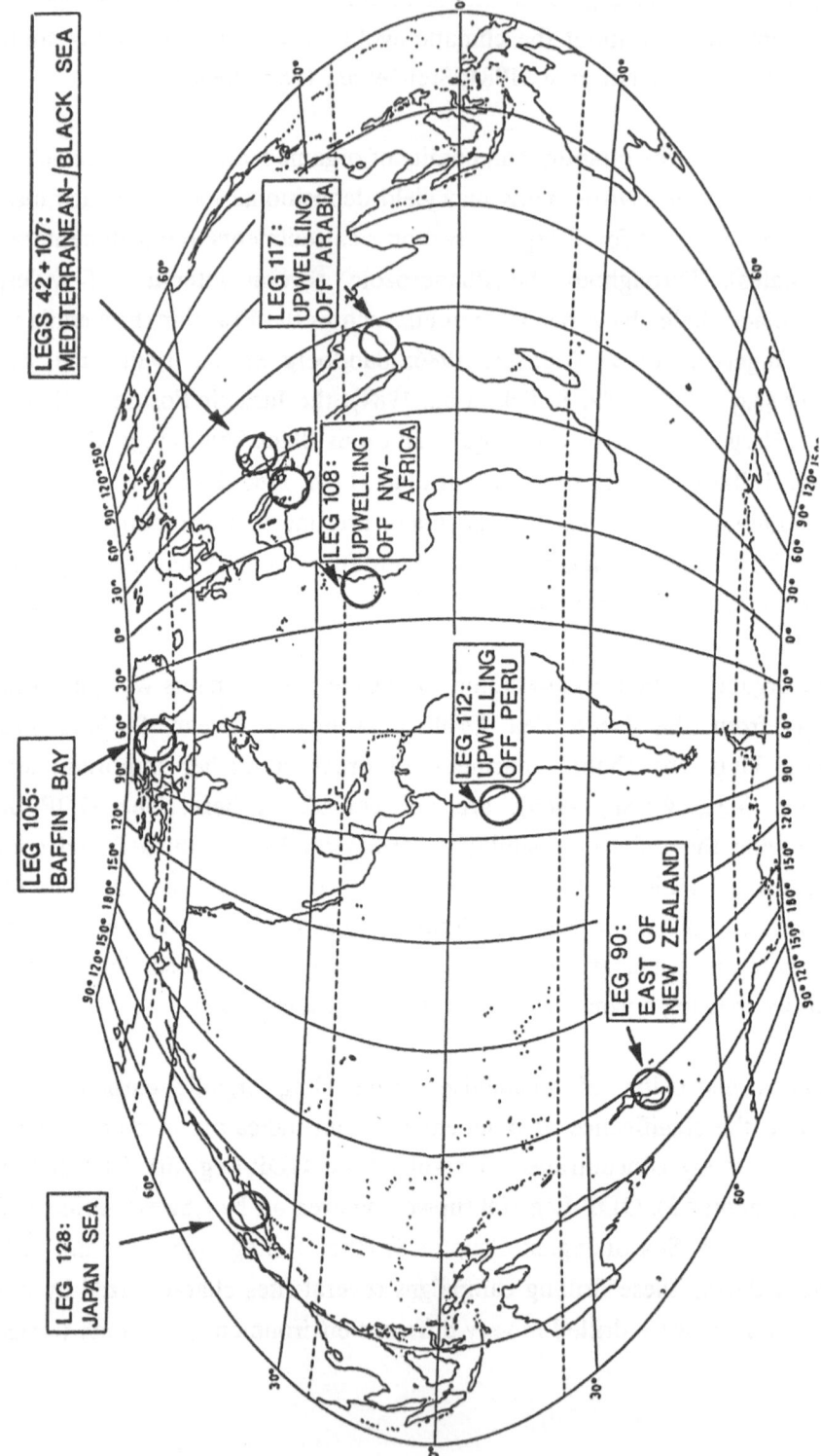

Fig. 2: DSDP/ODP Legs discussed in this study.

supply, upwelling, and anoxic environments, respectively; Fig. 3). In addition, results from investigations of sediments from DSDP-Legs 42 and 90 and ODP-Legs 107, 112, and 117 (Fig. 2) are discussed.

Major objectives of this study are (i) to reconstruct changes in paleoceanographic variables (i.e., surface and deep-water circulation patterns, upwelling intensity, surface-water productivity, and oxygen content of deep water) and the history of paleoclimate in these different environments through late Cenozoic time and (ii) to develop more general models for organic carbon accumulation from the results of the individual case studies. In a short epilogue, the data derived from investigations of Cenozoic organic-carbon-rich sediments are compared with data from Cretaceous Atlantic Ocean black shales (Chapter 10).

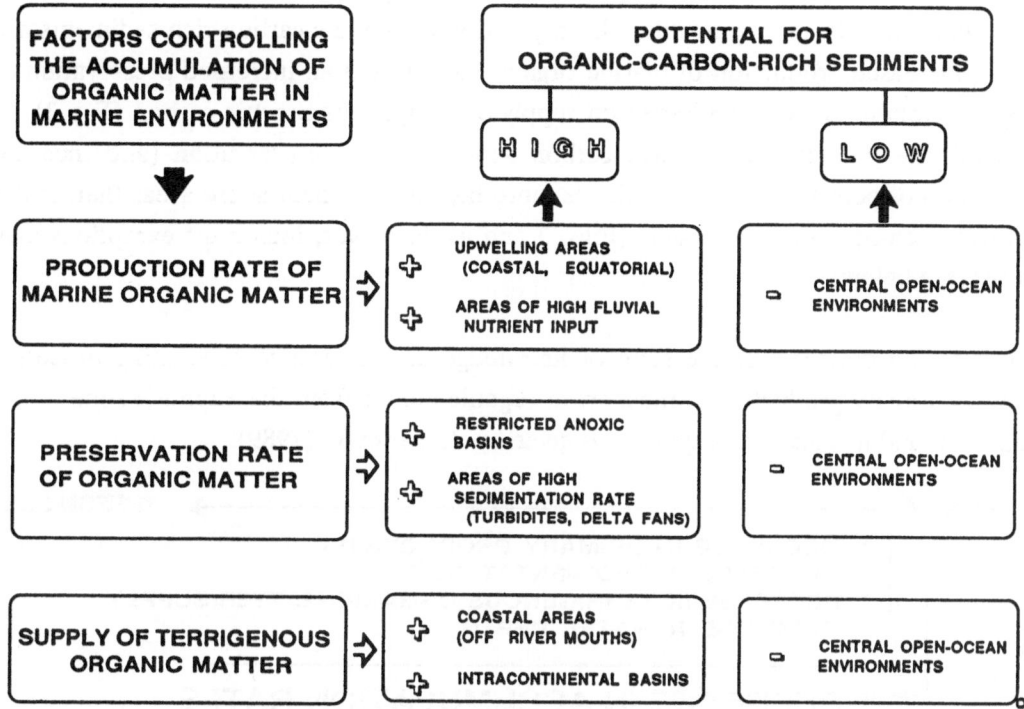

Fig. 3: Factors controlling the accumulation of organic matter in marine environments.

Since information about the composition of the organic matter is absolutely necessary for reconstruction of the depositional environment from organic carbon data, a further topic of this study is to compare organic carbon characteristics derived from different analytical techniques (Chapter 4). The use of different techniques for determination of organic matter composition is very useful because each of the standard methods has its advantages and disadvantages and the use of a single method may cause misleading interpretations.

2. Factors controlling organic carbon accumulation

Since "normal" open-ocean sediments are typically depleted rather than enriched in organic carbon and sediments with organic carbon contents of more than 1 % are largely restricted to minor parts of the modern ocean, special environmental conditions are required to allow the formation of these "unusual" organic-carbon-rich sediments. The most important factors controlling the deposition of organic-carbon-rich sediments are (i) an increased production of marine organic carbon, (ii) an increased preservation of organic carbon, and (iii) an increased supply of terrigenous organic carbon (Fig. 3). In general, the potential for organic carbon deposition and preservation (and thus the formation of petroleum source rocks) is more favorable in near-shore areas than in the central open-ocean environments (Figs. 3 and 4). However, important exceptions may occur (see below).

An excellent overview of the state of knowledge and ignorance concerning processes such as primary productivity, transport of organic matter from the euphotic zone to the sea floor, and its burial in sediments, is given by Berger et al. (1989).

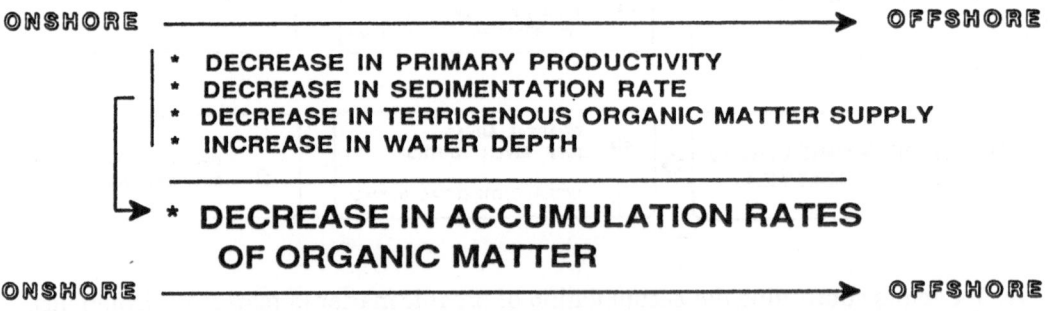

Fig. 4: Factors causing decrease in accumulation of organic matter in offshore environments.

2.1. Production of marine organic carbon and flux to the sea floor

In spite of differences in detail, the global distribution of primary productivity shown in Figure 5 (Koblents-Mishke et al., 1970), generally resembles that of the organic carbon distribution of surface sediments (Fig. 1), indicating the productivity to be one important factor for the formation of organic-carbon-rich sediments. From this productivity map it is obvious that the central ocean gyres are characterized by low productivity values of less than 50 gC m^{-2} yr^{-1}. On the other hand, the most productive regions are the coastal upwelling areas such as off Northwest and Southwest Africa, Arabia, and Peru, where upwelling brings nutrients to the surface (Fig. 5). In these upwelling areas, average productivity is around 250 gC m^{-2} yr^{-1} (Tab. 1). Occasionally, distinctly higher values have been reported (e.g., 600 gC m^{-2} y^{-1} off Peru; Suess, 1980). Organic carbon contents as high as 20 % were measured in sediments from the upwelling area off Peru (Müller and Suess, 1979). In marginal seas, such as the Sea of Japan (Fig. 5) and other coastal areas with a high fluvial nutrient supply (Schemainda et al., 1975), productivity values similar to those typical for upwelling areas may occur. In equatorial upwelling zones, productivity values of more than 100 gC m^{-2} yr^{-1} were recorded.

Tab. 1: Global net primary production (from Romankevich, 1984)

Ecosystem	Area (10^6 km^2)	10^9t C_{org} yr^{-1}	gC_{org} m^{-2} yr^{-1}
Estuaries	1.4	1.0	714.3
Upwelling zones	0.4	0.1	250.0
Continental shelf	26.6	4.3	161.6
Open ocean	332.0	18.7	56.3

From Table 1 it is obvious that the organic carbon produced in the high-productivity areas is only a minor proportion of the total annual production of organic carbon in the world ocean because these areas are very small in comparison to the open ocean area.

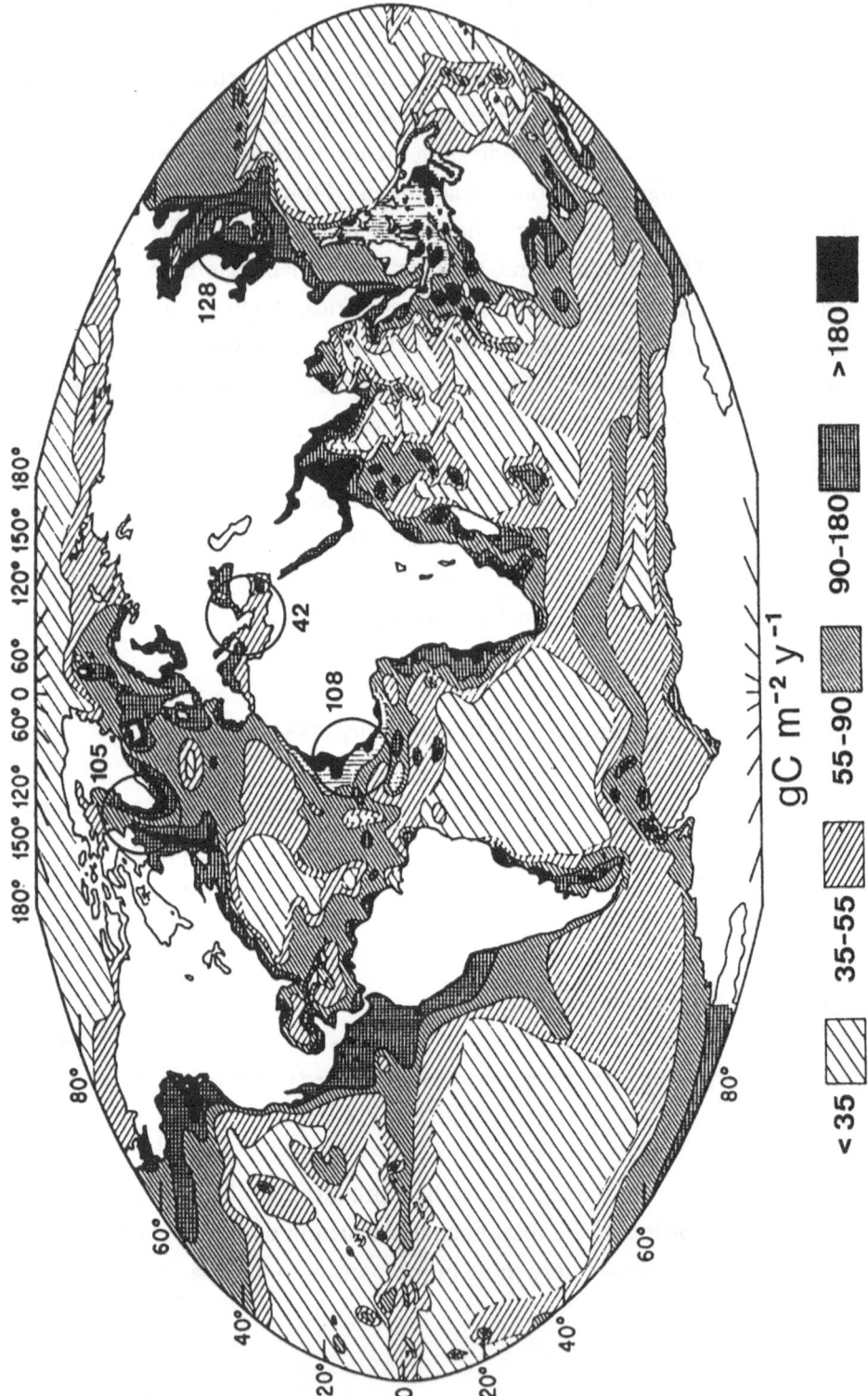

Fig. 5: Distribution of global surface-water productivity (from Mopper and Degens, 1976, according to Koblents-Mishke et al., 1970). Encircled areas are discussed in the text.

gC m^{-2} y^{-1}

< 35   35-55   55-90   90-180   >180

However, according to a compilation of Berger et al. (1989), one quarter of the total production occurs in slightly less than 10% of the ocean area, one half of the production in about 30% of the ocean area. These are mainly the shelf/slope environments. Furthermore, in these environments water depths are generally shallower and sedimentation rates higher than in the deep ocean, resulting in distinctly higher flux and preservation rates of organic carbon at the sea floor (see below). Thus, the shelf/ slope areas with their enhanced surface-water productivity are of major importance for the global carbon budget.

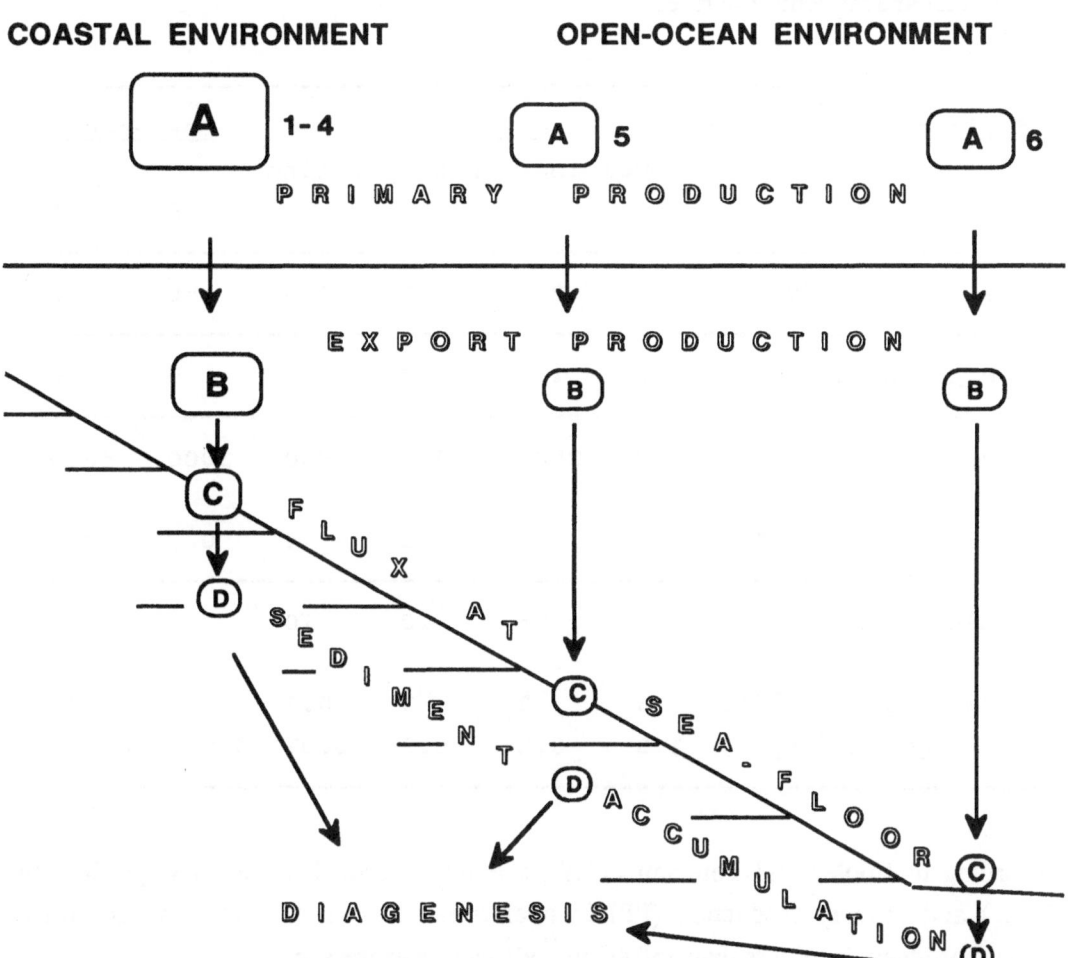

Fig. 6: Scheme indicating major steps of organic-carbon path from surface water into the deep-sea sediment. For letters A-D and numbers 1-6 see Table 2.

Only part of the organic matter produced in the euphotic zone leaves this zone as known from sediment traps installed immediately below the euphotic zone (Fig. 6). This flux of organic matter from the euphotic zone is the "export production" which, in this definition, is equivalent to the "new production" (Berger et al., 1989; Eppley, 1989).

Tab. 2: Examples of (A) surface-water productivities (PP) (after Romankevich 1984), (B) new production (P_{new}) estimated using equation (2), (C) carbon flux at the seafloor (CF_{sf}) at $depth_{sf}$ estimated using equation (3), and (D) carbon accumulation rates in surface sediments (CA) from different environments (cf. Fig. 6 for No. 1 to 6 and A to D); for further explanations see text.
LSR = Linear sedimentation rates

	Environment		Coastal				Open-ocean	
			Upwelling		Non-upwelling			
			1	2	3	4	5	6
A	PP	($gC\ m^{-2}y^{-1}$)	250	250	150	150	50	50
B	P_{new}	($gC\ m^{-2}y^{-1}$)	110	110	50	50	5	5
	$Depth_{sf}$	(m)	250	2500	250	2500	3000	5000
C	CF_{sf}	($gC\ m^{-2}y^{-1}$)	30	7	15	4	0.7	0.5
	LSR	($cm\ ky^{-1}$)	20	15	10	10	5	1
D	CA	($gC\ m^{-2}y^{-1}$)	13	2	3	0.7	0.07	0.01
		($gC\ cm^{-2}\ ky^{-1}$)	1.3	0.2	0.3	0.07	0.007	0.001

According to Eppley and Peterson (1979), the relationship between new production (P_{new}) and primary production (PP) in open-ocean to coastal environments with a primary productivity of less than 150 $gC\ m^{-2}\ y^{-1}$ can be expressed as

$$(1) \qquad P_{new}/PP = PP/400$$

As a refinement of this, Berger et al. (1989) suggest an equation in the form of

$$(2) \quad P_{new}/PP = PP/400 - PP^2/340000$$

For productivity values between 0 and 500 gC m^{-2} y^{-1}, this equation provides an excellent fit to the data given by Eppley and Petersen (Berger et al., 1989).

During its fall through the oxic water column, the organic matter continues to decay quickly as has been shown in sediment trap studies. Only a small amount of the primary organic matter reaches the sea floor (Fig. 6, Tab. 2; Suess, 1980; Betzer et al., 1984; Walsh, 1989). According to Betzer et al. (1984), this downward flux of organic carbon can be estimated as

$$(3) \quad CF = 0.409 \, PP^{1.41} \, DEP^{-0.63}$$

where CF (in gC m^{-2} y^{-1}) is the flux of organic carbon at depth DEP (in m) and PP (in gC m^{-2} y^{-1}) the primary productivity at the surface. Using DEP = water depth of the sea floor, it is possible to estimate the amount of primary organic carbon reaching the sea floor (see examples in Tab. 2). That means, that only this portion of organic carbon is the maximum which is available to be buried into the sediment. How much of this survives further decomposition processes at the sea floor or during earliest diagenesis and is finally preserved in near-surface sediments, depends on additional factors such as the sedimentation rate and the oxygen content of the bottom water (see Chapter 2.2.). The varying influence of these parameters on organic carbon accumulation in the marine environment makes estimates of paleoproductivity from sediment data so difficult (see Chapter 3; cf., Emerson, 1985; Emerson et al., 1985; Sarnthein et al., 1987).

There is one further point which makes estimates of (paleo) flux and (paleo) productivity even more complicated. Detailed measurements of surface-water productivity and sediment-trap experiments using time series traps in year-long deployments, have shown that significant seasonal and interannual variations in surface-water productivity and flux to the seafloor occur in most parts of the world ocean (e.g., Honjo, 1982; Postma, 1982; Deuser, 1986; Fischer et al., 1988; Wefer et al., 1988). Especially in high latitudes, seasonality of flux can produce extreme fluctuations, as shown, for example, in time series trap experiments in the Bransfield Strait (Antarctica) where particle flux is almost completely restricted to a short period of time during the (ice-free) southern summer (Fig. 7; Wefer et al., 1988). On the other hand, sediment data will generally reflect the average longer-term productivity record.

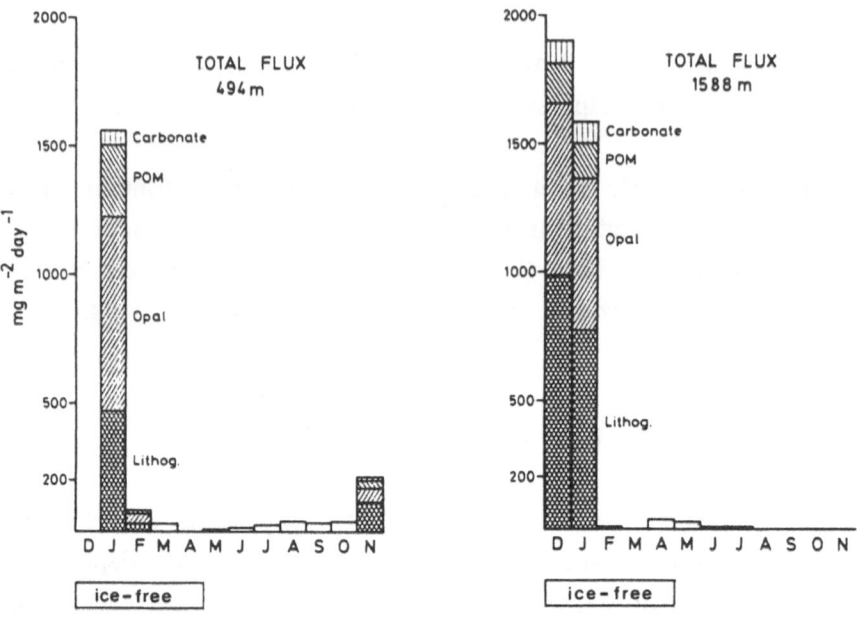

Fig. 7: Variability of particle flux into sediment traps: Seasonal flux at 499 m and 1588 m water depth in Bransfield Strait, Antarctica, between 1 December 1983 and 25 November 1984 (Wefer et al., 1988). POM = particulate organic matter.

2.2. Preservation of organic carbon

The preservation rate of organic carbon in the marine environment is controlled by several factors including oxygen content of bottom water, bulk sedimentation rate, extent of bioturbation, and composition of organic matter (e.g., Müller and Suess, 1979; Demaison and Moore, 1980; Aller and Mackin, 1984; Emerson, 1985; Calvert, 1983; Heggie et al., 1987; Canfield, 1989; Reimers, 1989).

In general, oxygen-depleted environments are characterized by sediments with high contents of (hydrogen-rich) organic matter (e.g., Demaison and Moore, 1980). They occur where the demand of oxygen in the water column exceeds the supply. These conditions can be developed in two different ways (e.g., Thiede and van Andel, 1977; Demaison and Moore, 1980):

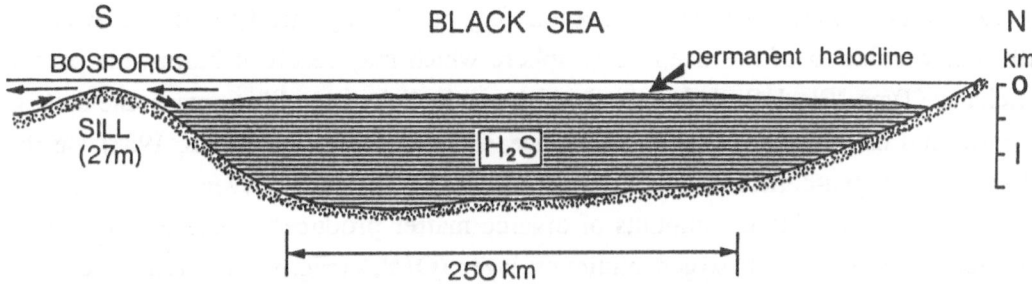

Fig. 8: Scheme indicating the anoxic depositional environment of the Black Sea (after Demaison and Moore, 1980).

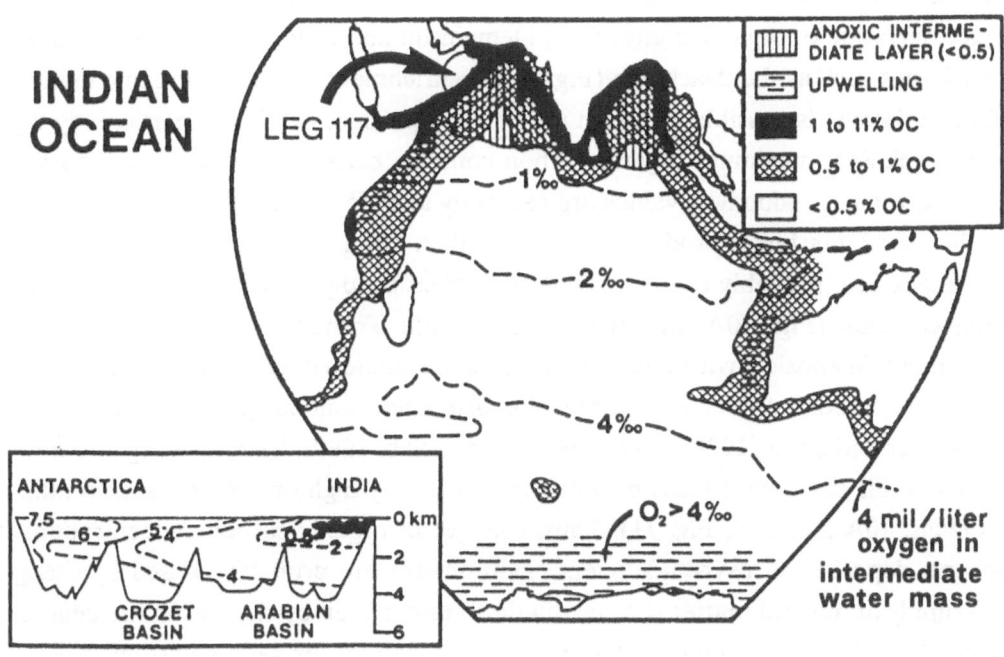

Fig. 9: Organic-carbon content in surface sediments and oxygen content of water masses in the Indian Ocean (from Demaison and Moore, 1980). Encircled area upwelling area drilled during ODP-Leg 117.

An extremely reduced vertical circulation (e.g., a density stratification) prevents a sufficient ventilation of the deep-water sphere which may result in basin-wide oxygen deficiency (Type "Black Sea"; Fig. 8). In high-productivity upwelling environments (e.g. off Peru and off Namibia; Calvert and Price, 1983; Reimers and Suess, 1983), on the other hand, oxygen-deficient conditions are caused by the high oxygen demand due to the decomposition of large amounts of organic matter produced in the euphotic zone and lead to an expanded oxygen minimum layer (OML) (Fig. 9). In areas where the OML impinges on the continental slope, the sediment/water interface and the overlying water masses may become strongly oxygen-depleted (cf. Fig. 132A, position 2).

The main difference between both possibilities is that in high-productivity environments the increased flux of marine organic matter causes dynamic oxygen depletion whereas in restricted anoxic basins stagnant deep-water conditions support increased preservation and accumulation of organic matter (e.g., Demainson and Moore, 1980). This means, in anoxic basins such as the Black Sea (Fig. 8), an enrichment of marine organic matter in surface sediments is possible although the primary productivity is not particularly high. In the Black Sea, maximum organic carbon contents occur in the center of the basin (Fig. 10A) where productivity values are relatively low (55-90 gC m^{-2} y^{-1}; Fig. 10B). On the other hand, maximum productivity of more than 500 gC m^{-2} y^{-1} in the coastal zone of the Black Sea, probably caused by fluvial nutrient supply, coincides with low organic carbon contents (Fig. 10A and 10B; Shimkus and Trimanis, 1974; Sorokin, 1982). Furthermore, in anoxic environments there is no correlation between sedimentation rate and organic carbon content as recorded for oxic environments (Fig. 11; Heath et al., 1977; Müller and Suess, 1979, Ibach, 1982; Stein, 1986b, 1990). Very high organic carbon contents in this case may be accompanied by either very high or very low sedimentation rates (Figs. 10A and 10C, Fig. 11). Thus, changes of organic carbon concentrations in sediments deposited in oxygen-deficient environments are probably caused by changes in the supply of mineral matter (i.e., by dilution), rather than by changes in the degree of preservation of marine organic matter due to variations of sedimentation rate (see below) or productivity. Changes in the supply of terrigenous organic carbon may also have an influence on organic carbon contents (see Chapter 2.3.).

A stable density stratification preventing the ventilation of bottom-water masses and resulting in suboxic to anoxic conditions at depth, is also typical of major parts of the modern Baltic Sea. In this environment, surface sediments with organic carbon contents of more than 5 % are common (Kögler and Larson, 1979; Müller and Suess, 1979; Rother, 1989; Niedermeyer and Lange, 1990; Schubert and Stein, unpubl.).

15

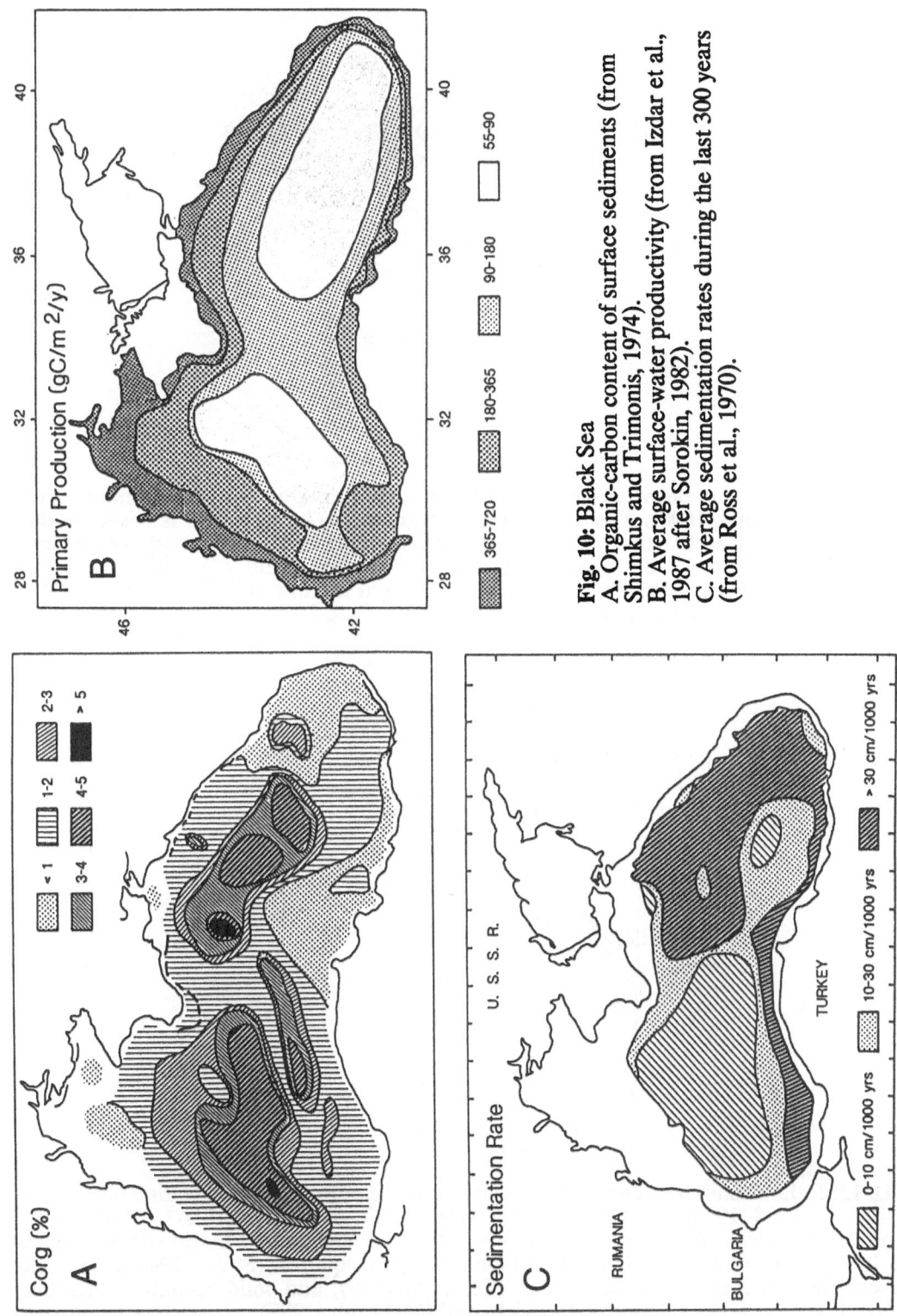

Fig. 10: Black Sea
A. Organic-carbon content of surface sediments (from Shimkus and Trimonis, 1974).
B. Average surface-water productivity (from Izdar et al., 1987 after Sorokin, 1982).
C. Average sedimentation rates during the last 300 years (from Ross et al., 1970).

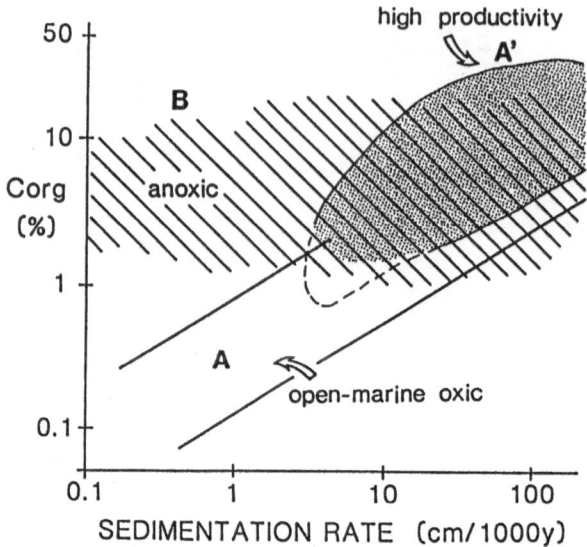

Fig. 11: Relationship between (marine) organic-carbon content and sedimentation rates (from Stein, 1986b, 1990). The distinction between fields A, A', and B is based on data derived from Recent to Miocene sediments deposited in normal open-ocean environments (field A), upwelling high-productivity areas (field A'), with the stippled area indicating coastal upwelling and the open area at the low end of field A' indicating equatorial upwelling data, and anoxic environments (field B).

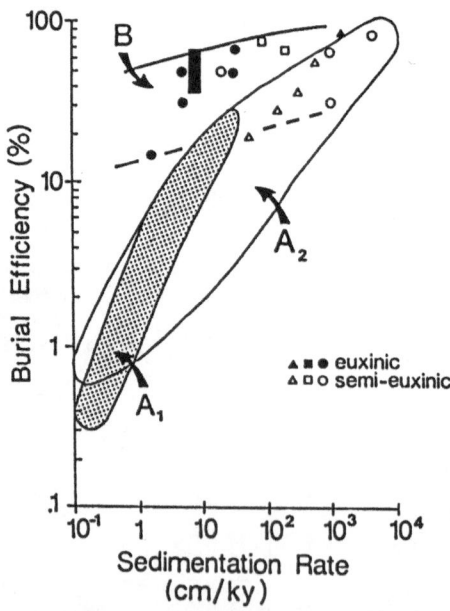

Fig. 12: Relationship between burial efficiency of organic carbon and sedimentation rate (after Canfield, 1989; Henrichs and Reeburgh, 1987; Stein, 1986a). A1 (Stein, 1986a) and A2 (Henrichs and Reeburgh, 1987) = data from normal oxic marine environments. B = Euxinic to semi-euxinic environments: Open squares (Stein, 1986a); triangles (Henrichs and Reeburgh, 1987); dots (Canfield, 1989); and solid rectangle data from Vaynshteyn et al. (1986) cited in Henrichs and Reeburgh (1987).

In oxic environments, respiration is the main mechanism decomposing organic matter whereas in anoxic (deep-water) environments sulfate reduction becomes the most important mechanism. For semi-anoxic environments, both oxic respiration and sulfate reduction will decompose organic matter (Froelich et al., 1979; Henrichs and Reeburgh, 1987; Canfield, 1989). In recent studies, Henrichs and Reeburgh (1987) and Canfield (1989) have shown that rates of organic matter decomposition in anoxic and semi-anoxic sediments can be similar to those found in normal marine sediment at the same sedimentation rate.

According to Henrichs and Reeburgh (1987), the efficiency of carbon burial (i.e., the ratio of organic carbon accumulation rate in surface sediments and flux of organic carbon at the sea floor) is related to the bulk sedimentation rate. The burial efficiency shows no clear difference between oxic and anoxic (euxinic) environments (Fig. 12). A similar relationship has been shown for oxic environments by Stein (1986a), based on data from Müller and Suess (1979) and the carbon-flux equation of Betzer et al. (1984) (Fig. 12). Canfield (1989) added data from anoxic environments characterized by relatively low sedimentation rates and indicated that at lower sedimentation rates burial efficiencies are higher for anoxic sediments than for oxic sediments of similar sedimentation rates (Fig. 12). This agrees well with new data from the western part of the Black Sea where efficiencies of 33 to 70% were determined for sediments with sedimentation rates of 5 to 10 cm/ky (Fig. 12; Vaynshteyn et al. 1986, cited in Henrichs and Reeburgh, 1987).

This conclusion is different from that of Henrichs and Reeburgh (1987) (see above). It has to be kept in mind, however, that in the study of Henrichs and Reeburgh (1987) most of the data are from normal marine environments, and all data from anoxic environments are from areas with high sedimentation rates and high productivity, i.e., all data falling into fields A and A' of Figure 11 (cf., Stein, 1986b, 1990). A combination of the data from anoxic (and semi-anoxic) environments of Vaynshteyn et al. (1986), Stein (1986a), Henrichs and Reeburgh (1987) and Canfield (1989), leads to a different relationship between carbon burial efficiency and sedimentation rate which is not as clear as that found for oxic environment (Fig. 12). The distribution of data in Figure 12 resembles that in Figure 11 and yields similar interpretations (cf., Stein, 1986b; 1990). In areas of high sedimentation rates, burial efficiency is similar in oxic and anoxic enviroments, but, burial efficiency and preservation of organic carbon are significantly higher in anoxic deep-water environments with low sedimentation rates.

Furthermore, anaerobic degradation results in a more lipid-hydrogen-rich organic matter than that remaining after aerobic degradation (Pelet and Debyser, 1977; Demaison and Moore, 1980). Thus, in conclusion, sediments characterized by high amounts of hydrogen-rich organic carbon and low sedimentation rates strongly suggest deposition under anoxic conditions with increased preservation rate of organic carbon.

An additional example of increased preservation of organic carbon is shown by investigations of turbidites from the African continental margin. In these turbidites intercalated in organic-carbon-lean sediments, organic carbon of more than 1 % is preserved in oxic deep-sea environments (Weaver and Kuijpers, 1983; Colley et al., 1984). Since the organic matter in turbidites and slumps is rapidly buried after redeposition in the deep sea, the residence time in zones of bioturbation and oxic decomposition is short. In this way, elevated quantities of (hydrogen-rich) organic matter can also be preserved in an oxygen-rich open-marine environment (cf. Cornford et al., 1979; Jones, 1983; Arthur et al., 1984).

2.3. Supply of terrigenous organic carbon

The annual supply of terrigenous organic carbon to the world ocean is small in comparison to the amount of organic carbon produced by marine organisms (Romankevich, 1984). Most of the (riverine) terrigenous organic matter is accumulated in coastal environments (e.g., Degens and Mopper, 1976). Furthermore, about 35 % of the riverine organic matter (about 81×10^6t C_{org} y^{-1}) belongs to the labile (metabolizable) fraction and may become oxidized already in estuaries and near-shore environments, whereas the other 65% (about 150×10^6t C_{org} y^{-1}) appears to be highly refractory and may be transported further offshore (Ittekott, 1988). In estuaries, such as the St. Lawrence Estuary, sediments may contain up to about 3 % terrigenous organic carbon (Fig. 13; Tan and Strain, 1979). However, already in the Gulf of St. Lawrence where similar organic carbon values were determined in surface sediments, a major portion is of marine origin as deduced from $\delta^{13}C_{org}$ data (Fig. 13) and C/N ratios (Pocklington, 1976). This is mainly caused by increased surface-water productivity in these coastal environments due to fluvial nutrient supply (cf., Postma, 1982; Romankevich, 1984).

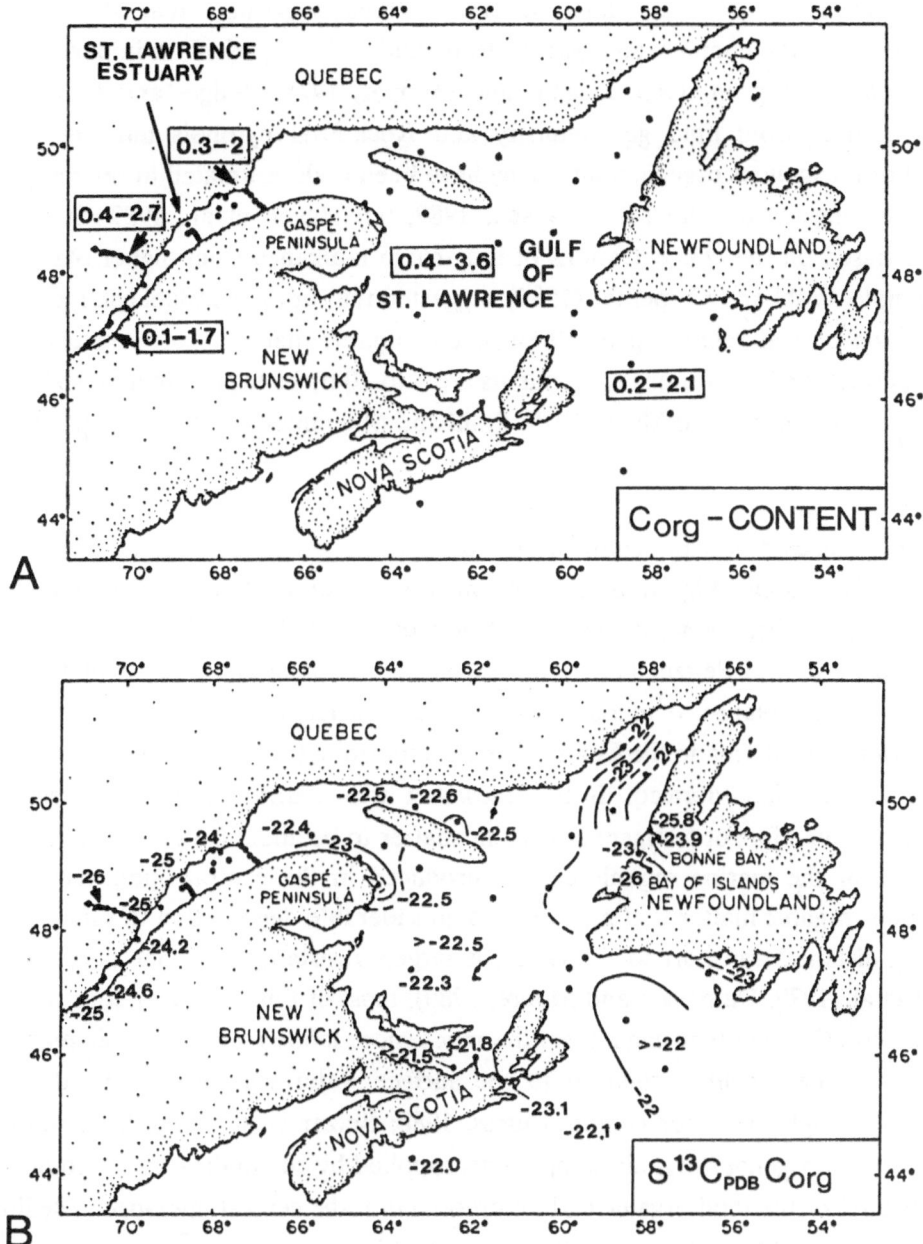

Fig. 13: (A) Organic carbon content (%) in surface sediments of the St. Lawrence Estuary and the Gulf of St. Lawrence and (B) carbon-stable isotopes of organic carbon ($^0/_{00}$PDB) (Tan and Strain, 1979).

Mainly based on $\delta^{13}C_{org}$ data, several authors suggested that the terrestrial contributions to the open-ocean organic carbon budget is only of minor importance (e.g., Sackett and Thompson, 1963; Degens and Mopper, 1976; Hedges and Mann, 1979). However, data from lipid geochemistry and Rock-Eval pyrolysis indicate that the amount of terrigenous organic matter may have been underestimated in several parts of the ocean (e.g., Prahl and Muehlhausen, 1989; ten Haven et al., 1989; Stein et al., 1989b). Based on the concentrations of long-chain n-alkanes from higher-plant waxes, for example, Prahl and Carpenter (1984) suggested that in the hemipelagic muds of the Washington coastal region up to 30 % of the organic matter is of terrestrial origin. These estimates of terrigenous components are 5 to 10 times higher than those based on $\delta^{13}C_{org}$ data (Hedges and Mann, 1979; Prahl and Muehlhausen, 1989; see also Chapter 4).

In the organic-carbon-lean sediments of the central open-ocean regions, terrigenous organic carbon supplied by winds may become a major proportion of the organic matter accumulated in these environments (Zafiriou et al., 1985; Prahl and Muehlhausen, 1989). Gagosian and Peltzer (1987) measured high relative abundances of long-chain series of vascular plant-wax lipids (i.e., n-alkanes, n-acids, n-alkohols) in aerosols from the Enewetak Atoll area in the Central Pacific, indicating that significant amounts of the organic matter is land-derived. In the Enewetak area, Zafiriou et al. (1985) estimated the annual eolian flux of terrigenous organic matter to be about 0.5 gC m^{-2} y^{-1}, which is about 1% of the marine organic carbon produced in the central open ocean. Since terrigenous organic matter that has survived considerable degradation in subaerial soils and long-distance transport to the ocean is largely resistant to further decomposition (Tissot et al., 1979; Demaison and Moore, 1980), most of this organic matter will reach the sea floor. On the other hand, from the much more labile marine organic matter less than 1% of the amount produced in the surface water will survive the downward transport through the water column (Suess, 1980; Betzer et al., 1984; Emerson et al., 1987). Thus, terrigenous organic components supplied by eolian dust may account for as much as 50% of the total organic carbon preserved in the central open-ocean sediments (Zafiriou et al., 1985; Prahl and Muehlhausen, 1989).

3. Reconstruction of depositional environments from sediment data

The main objective of this synthesis is the reconstruction of depositional environments in which organic-carbon-rich sediments could be formed. Thus, it is of major interest to get information from sediment data about (1) the paleoproductivity of surface water masses, (2) the oxygen content of deep water masses and its influence on organic carbon preservation, and (3) the supply of terrigenous organic carbon as a function of paleoclimate.

3.1. Estimates of paleoproductivity

Indicators useful for estimating surface-water paleoproductivity from sediment data are summarized in Figure 14 (cf., Bruland et al., 1989):

- Flux rates of biogenic compounds such as marine organic carbon, biogenic silica, and biogenic carbonate may provide information on surface-water productivity (cf., Berger et al., 1989 and references therein). It has to be considered, however, that the latter two components may be strongly influenced by dissolution effects whereas marine organic carbon is less resistent towards decomposition. Furthermore, when organic carbon data are beying used, it is also of major importance to subtract the terrigenous proportion before interpreting the data in terms of productivity (see Chapter 4.).

```
            PROXIES FOR
      SURFACE-WATER PRODUCTIVITY

  •••  FLUX RATES OF:
         - MARINE ORGANIC CARBON
         - BIOGENIC OPAL  (DIATOMS,  RADIOLARIANS)
         - CALCIUM CARBONATE  (FORAMINIFERS, COCCOLITHS)
         - PHOSPHATES

  •••  ORGANIC CARBON / SEDIMENTATION RATE RELATIONSHIP

  •••  FAUNAL AND FLORAL ASSEMBLAGES

  •••  CARBON AND NITROGEN STABLE ISOTOPES

  •••  BIOMARKERS

  •••  TRACE METALS  (Cd, Ba, Cu)
```

Fig. 14: Proxies of surface-water productivity in sediment data.

- In an oxic environment, the relationship between organic carbon content and sedimentation rate has been used to get rough estimates of paleoproductivity (Fig. 11; Müller and Suess, 1979; Stein, 1986b, 1990). In central open-ocean low-productivity environments, low organic carbon contents (< 0.4%) and low sedimentation rates (0.2-1 cm/1000 years) are typical, whereas in high-productivity upwelling environments very high organic carbon contents (up to 20%) and very high sedimentation rates occur (up to some 100 cm/1000 years).

- In order to get more quantitative estimates of paleoproductivity from organic carbon data, Müller and Suess (1979) derived an empirical formula which is based on the relationship between organic carbon accumulation rates of surface sediments, bulk sedimentation rates, and measured (i.e., Recent) surface-water productivity data. A combination of the results of Müller and Suess (1979) and the results on organic-carbon flux to the seafloor of Betzer et al. (1984), led to the following equation which is used in this study for estimates of paleoproductivity (for detailed deduction and limits of use of the equation, see Stein, 1986a):

$$(4) \quad PP = 5.31 \, (C \, (WBD - 1.026 \, PO/100))^{0.71} \, LSR^{0.07} \, DEP^{0.45}$$

where PP ($gC \, m^{-2} \, y^{-1}$) is the primary productivity, C (%) the (marine) organic carbon content, WBD ($g \, cm^{-3}$) the wet bulk density, PO (%) the porosity, LSR ($cm \, ky^{-1}$), and DEP (m) the water depth.

A similar equation for estimating paleoproductivity, also based on the original work of Müller and Suess (1979), was presented by Sarnthein et al. (1987). Productivity values calculated with this equation and equation (4) are shown to be linearly related but slightly different in absolute values (Stein et al., 1989b). In 1988, Sarnthein et al. published another version of the paleoproductivity equation for estimating new production rates (cf. Chapter 2.1.).

When using the equations of Müller and Suess (1979), Stein (1986a) or Sarnthein et al. (1987, 1988) it has to be considered that they describe the relationship between surface-water productivity and organic carbon accumulation in oxic environments. This means these empirical equations cannot be used to estimate productivities in anoxic environments such as those typical of the modern Black Sea or the Cenomanian/Turonian North Atlantic (cf., Stein, 1986a). For anoxic environments,

Bralower and Thierstein (1984), from comparison of accumulation rates of (marine) organic carbon and primary production rates in recent anoxic environments, suggested that at least 2% of the primary organic carbon is preserved in the sediment. Based on this, they derived the following equation for estimating paleoproductivities in anoxic environments:

$$(5) \quad PP = 5 \ C \ LSR \ (WBD - 1.026 \ PO/100)$$

where PP (in $gC \ m^{-2} \ y^{-1}$) is the primary productivity, C (in %) the (marine) organic carbon content of the sediment, WBD (in $g \ cm^{-3}$) the wet bulk density, and PO (in %) the porosity. This formula is has been used by Stein (1986a) for calculating the paleoproductivities in the Cenomanian/Turonian Atlantic.

A more detailed discussion of the assessment of paleoproductivity from organic carbon data using empirical equations was presented by Berger et al. (1989).

- Faunal and floral assemblages may reflect surface-water productivity, too (e.g., Berger and Killingley, 1977; Thiede, 1983; Berger et al., 1989 and further references therein). Mix (1989), for example, used planktonic foraminifera assemblages and standard transfer function techniques (Imbrie and Kipp, 1971) for quantitative estimates of productivity in the Atlantic Ocean. Upwelling productivity can also induce specific assemblages of benthic foraminifera as shown by Lutze and Coulbourn (1984), Altenbach (1985), and Altenbach and Sarnthein (1989).

- $\delta^{13}C$ values of benthic foraminifera, as a measure of total dissolved CO_2 in the deep water, may give information about local surface ocean fertility (e.g., Duplessy et al., 1984; Shackleton and Pisias, 1985; Vincent and Berger, 1985; Zahn et al., 1986). When interpreting $\delta^{13}C$ data it has to be considered, however, that the amount and composition of CO_2 in the deep water are controlled by two factors, (i) the local surface-water productivity and the vertical flux of organic carbon and (ii) the global renewal rate of oxygenated deep water. The use of $\delta^{13}C$ values determined on both infaunal species (such as *Uvigerina peregrina*) and epibenthic species (such as *Cibicidoides wuellerstorfi*) may help to distinguish between both factors (Duplessy et al., 1984; Zahn et al., 1986).

- Biomarkers such as long-chain alkenones can be used as paleoproductivity indicator (Volkman et al., 1980; Brassell and Eglinton, 1983; Marlowe et al., 1984; Brassell et al.,

1986). These alkenones are restricted to a narrow range of algae belonging to the class *Prymnesiophyceae*, of which *Emiliania huxleyi* is the most important member today. Thus, changes in alkenone concentrations may give a record of paleoproductivity for an important group of marine algae. It still has to be demonstrated, however, whether variations in the concentration of alkenones although they are considered relatively stable organic compounds, can also reflect changes in preservation efficiency within the water column and/or the sediment (Prahl and Muehlhausen, 1989; Prahl et al., 1989).

- Trace metals (such as Ba and Cd) are proxies of the nutrient content of the ocean (e.g., Boyle, 1986; Schmitz, 1987; Bishop, 1988; Lea and Boyle, 1990). Barium, for example, is distinctly enriched in sediments deposited underneath high-productivity zones. This enrichment is probably due to inorganic precipitation of barite within microenvironments (broken diatoms, fecal pellets, etc.) in the presence of decaying organic matter and silica surfaces (Bishop, 1988; Bruland et al., 1989). As shown in a very recent study of sediments from the upwelling area off Peru (von Breymann et al., 1990), however, barium as paleoproductivity indicator should be used with caution in areas of varying water depths and in reducing environments.

When using the tracers of paleoproductivity summarized in Figure 14, it has to be remembered that none of them is completely resistant to changes (e.g., dissolution or decomposition) during transport through the water column, at the sea floor, or in the sediment (Jumars et al., 1989). This means, the sediment data of a specific variable cannot be taken as a direct measure of its production rate in the surface water. Thus, further information about environmental conditions such as, for example, degree of dissolution, oxygen content in the water body or extent of bioturbation, is necessary (see above). From this it is obvious that different independant tracers should be used for reconstruction of paleoproductivity in order to minimize errors in interpretations.

3.2. Characteristics of oxic and anoxic environments

Informations about the oxygen content of deep/bottom water masses can be derived from different parameters (Fig. 15):

- The preservation of high amounts of marine organic carbon in deep-sea sediments in combination with low sedimentation rates may indicate anoxic deep water conditions.

Under these conditions, there is no correlation between organic carbon content and sedimentation rate at all (Fig. 11; Stein, 1986b, 1990).

- According to Leventhal (1983) and Berner (1984, 1989), the relationship between organic carbon content and (pyritic) sulfur may also help to distinguish oxic and anoxic environments (Fig. 16). In siliciclastic normal marine sediments, sulfur is mainly bound to pyrite (Berner, 1984). Sulfur is available in excess as sulfate in sea water and iron is available from silicates (clay minerals) and crytalline oxid phases (Goldhaber and Kaplan, 1974; Raiswell and Berner, 1987). The limiting factor for pyrite formation under normal oxic deep-water conditions is the amount of organic matter controlling the formation of reducing conditions in the near-surface sediments. In this environment, there is a positive correlation between (pyritic) sulfur and organic carbon (Fig. 16; Berner, 1984; Raiswell and Berner, 1987). Under anoxic deep-water conditions, H_2S already exists in the sea water. Thus, framboidal pyrite can already be formed in the water column, resulting in an excess of sulfur in the organic carbon/sulfur diagram (Fig. 16; Leventhal, 1983).

PROXIES FOR
OXIC - SUBOXIC - ANOXIC
BOTTOM-WATER CONDITIONS

*** AMOUNT AND COMPOSITION OF ORGANIC CARBON

*** ORGANIC CARBON / SULPHUR RELATIONSHIP

*** ORGANIC CARBON / SEDIMENTATION RATE RELATIONSHIP

*** CARBON STABLE ISOTOPES

*** TRACE- METALS

*** TYPE AND DEGREE OF BIOTURBATION

*** FAUNAL ASSEMBLAGES

Fig. 15: Proxies for oxic-suboxic-anoxic deep/bottom-water conditions in marine sediments.

- In modern marine environments, low dissolved oxygen concentrations are associated with low $\delta^{13}C$ values of benthic foraminifera (e.g., Curry and Lohmann, 1983, 1985).

- Distinct enrichments of heavy metals (such as Mo, Zn, V, Cr, Cu) may point to anoxic deep-water conditions and low sedimentation rates (Brumsack, 1980, 1986).

- The type and degree of bioturbation as well as the composition of faunal assemblages may allow a characterization of the depositional environment in terms of oxygen content of the deep water (e.g., Wetzel, 1983; Ekdale, 1989). The lack of any bioturbation resulting in finely laminated sediments strongly suggests true anoxic deep-water conditions (i.e., < 0.1 ml O_2/l H_2O).

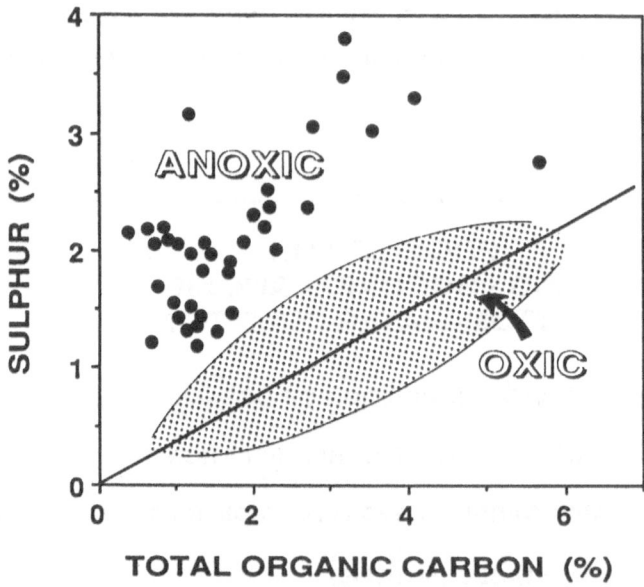

Fig. 16: Relationship between organic carbon content and (pyritic) sulfur in marine environments. Distinction between oxic and anoxic environment according to Leventhal (1983) and Berner (1984). For Quaternary normal marine, fine-grained detrital sediments, mean C/S weight ratio = 2.8 (Berner, 1984; Goldhaber and Kaplan, 1974); envelope of regression line encloses several hundred data points (Berner, 1989). Solid large dots are data from Black Sea sediments.

3.3. Terrigenous sediment supply and its paleoenvironmental significance

The characteristics of the terrigenous sediment fraction in marine sediments may provide important information on the climatic evolution of the past (Fig. 17):

- Using oxygen stable isotopes, changes in terrigenous sediment supply can be correlated with the global climate record (e.g., Shackleton et al., 1984; Stein, 1985a; Ruddiman and Janecek, 1989; Tiedemann et al., 1989; Jansen et al., 1990).

- Flux rates, composition, and grain size of the siliciclastic material (i.e., quartz, feldspars, and clay minerals) may allow (1) to distinguish different source areas, (2) to distinguish different transport mechanisms (i.e., fluvial vs. eolian vs. ice-rafting), and (3) to reconstruct the climate of the source area (e.g., Thiede, 1979; Sarnthein et al., 1981, 1982; Rea and Janecek, 1982; Stein, 1984, 1985a, 1985b, 1986c; Chamley, 1989; Tiedemann et al., 1989).

PROXIES FOR PALEOCLIMATE
IN MARINE SEDIMENTS

*** OXYGEN STABLE ISOTOPES

*** FLUX RATES, COMPOSITION, AND GRAIN SIZE
     OF SILICICLASTIC SEDIMENT FRACTIONS
     (QUARTZ, FELDSPARS, CLAY MINERALS)

*** FLUX RATES OF TERRIGENOUS ORGANIC MATTER

*** AMOUNT AND COMPOSITION OF POLLEN AND SPORES

*** PRESENCE OF FRESH-WATER DIATOMS

Fig. 17: Proxies for paleoclimate in marine sediments.

- Since terrigenous organic matter that has survived long-distance transport to the ocean is relative resistant to oxic decomposition (Tissot et al., 1979; Demaison and Moore, 1980), changes in accumulation rates of terrigenous organic carbon are primarily controlled by changes in rates of supply. Consequently, accumulation rates of terrigenous organic carbon may reveal information concerning the paleoclimate of the adjacent continent. High accumulation rates may imply intense fluvial runoff and a dense vegetation cover in the source area due to humid climatic conditions. On the other hand, during times of arid climatic conditions, reduced vegetation and the lack of river discharge may result in decreased accumulation rates of terrigenous organic carbon.

- The amount and composition of pollen and spores and the presence/absence of fresh-water diatoms in marine sediments are further important indicators for paleoclimatic conditions in the source area (e.g., Burckle and Akiba, 1977; Agwu and Beug, 1982; Dupont et al., 1989; Stabell, 1989).

4. Methods

4.1. Determination of the composition of organic matter

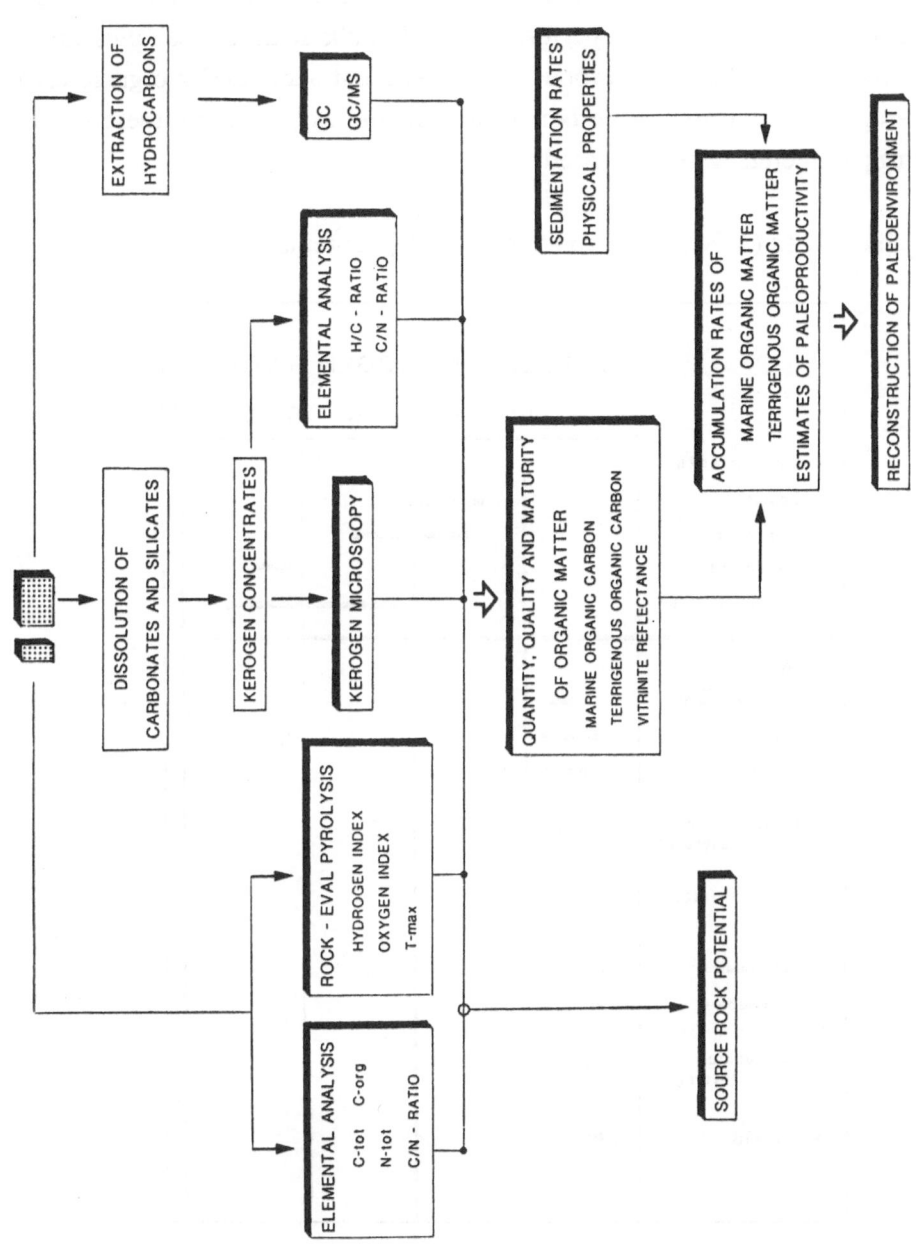

Fig. 18: Flow chart indicating methods used in this study for reconstruction of depositional environments from organic carbon data.

All methods used and discussed in this study are shown in the flow chart of Figure 18.

(1) Elemental analysis

Total and organic carbon, nitrogen, hydrogen, oxygen, and sulfur were determined using a HERAEUS CHN-Analyser with a supplement unit for oxygen and sulfur measurements. The analyses were performed on (i) bulk sediment samples for total carbon, nitrogen, and sulfur, (ii) carbonate-free sediment samples for organic carbon, and (iii) kerogen concentrates for carbon, hydrogen, and oxygen. From these raw data, the following parameters were calculated:

- Total organic carbon (TOC) and carbonate (CaCO₃) contents.

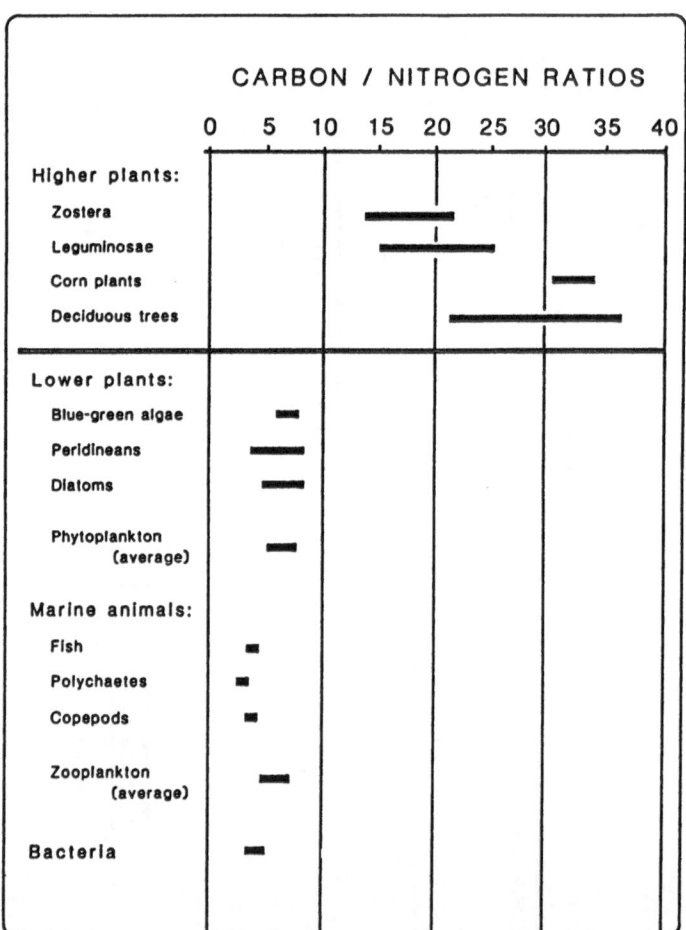

Fig. 19: Carbon/nitrogen (C/N) ratios of marine and terrestrial organisms (after Bordowskiy, 1965a, 1965b; Scheffer and Schachtschabel, 1984).

- Total organic carbon to total nitrogen (C/N) ratios. These ratios can be used to characterize the type of organic matter because different groups of organisms produce organic matter that contain different carbon and nitrogen contents. The mean C/N ratios for marine zoo- and phytoplankton is about 6 (Fig. 19) which may increase to values around 10 in organic detritus passing though the water column because of preferential decomposition of protein-rich components (e.g., Emery and Uchupy, 1984). Higher land plants, on the other hand, are characterized by C/N ratios of more than 15 (Fig. 19; Bordowskiy, 1965a, 1965b; Scheffer and Schachtschabel, 1984). When using C/N ratios it has to be taken into account, however, that in organic-carbon-poor sediments the amount of inorganic nitrogen (fixed as ammonium ions in the interlayers of clay minerals) may become a major portion of the total nitrogen (e.g., Stevenson and Cheng, 1972; Müller, 1977), causing (too) low C/N ratios. Therefore, in this study the discussion and interpretation of C/N ratios is restricted to organic-carbon-rich sediments, in which the nitrogen is predominantly organic.

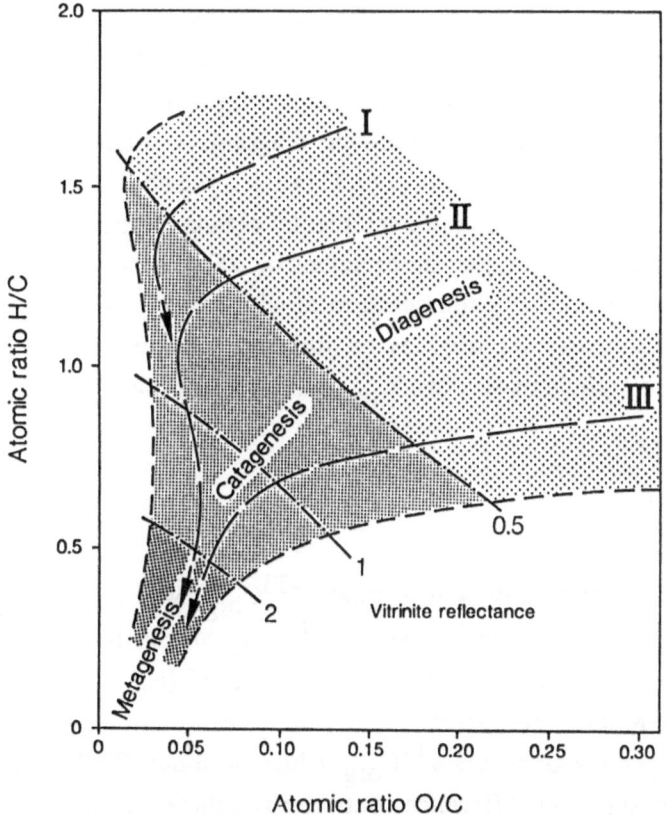

Fig. 20: Atomic hydrogen/carbon (H/C) and oxygen/carbon (O/C) ratios of organic matter (kerogen) from different source and different stage of maturity (from Tissot and Welte, 1984). Kerogen-types I and II = marine; kerogen-type III = terrigenous.

- Atomic hydrogen to carbon (H/C) and oxygen to carbon (O/C) ratios of kerogen concentrates. Plotted in a "van Krevelen diagram" (i.e., H/C vs. O/C), immature organic matter can be classified into "kerogen types I to III" (Fig. 20; Tissot and Welte, 1984). Marine organic carbon (type I and II) is characterized by high H/C ratios of about 1.2 to 1.5 and low O/C ratios around 0.1, whereas terrigenous organic matter (type III) has H/C ratios of less than 1 and O/C ratios around 0.2. With increasing maturity, H/C and O/C ratios decrease (Fig. 20). Thus, the H/C vs. O/C diagram may also give information about the level of thermal maturity. All data presented and discussed in this study fall into the diagenesis field of Figure 20, i.e., vitrinite reflectance values are less than 0.5 % due to the low thermal stress the organic matter has been exposed to.

(2) Rock-Eval pyrolysis

Rock-Eval pyrolysis was performed on bulk sediment samples and kerogen concentrates according to Espitalié et al. (1977). Hydrogen and oxygen contents of the samples, measured as hydrocarbon-type compound and carbon dioxide yields, respectively, were normalized to organic carbon and displayed as hydrogen index (mgHC/gC) and oxygen index (mgCO$_2$/gC). In a van Krevelen-type diagram, a classification of the organic carbon similar to that based on H/C and O/C ratios is possible (Figs. 20 and 41; cf., Orr, 1981; Dean et al., 1986). In immature sediments, organic matter dominated by marine components typically has hydrogen index values of 200 to 400 mgHC/gC (e.g., Stein et al., 1989a, b). Furthermore, the temperature at which pyrolysis yields the maximum of hydrocarbons (T_{max}), can be used as an indicator of the thermal maturity of the kerogen. Immature organic matter has T_{max} values of less than 435 °C.

(3) Stable carbon isotopes

The isotopic composition of organic carbon ($\delta^{13}C_{org}$) was measured by means of a Finnigan Delta mass spectrometer. These $\delta^{13}C_{org}$ values have long been used to characterize the nature of the organic source material (e.g., Sackett and Thompson, 1963; Hedges and Mann, 1979; Müller et al., 1983; Jasper and Gagosian, 1989). Marine plankton of temperate regions has $\delta^{13}C_{org}$ values of about -20 $^o/_{oo}$ (relative to PDB standard) with no apparent differences between zooplankton and phytoplankton (Fig. 21; Fontugne and Duplessy, 1981; Waples, 1981). However, the considerable variability of $\delta^{13}C_{org}$ among the individual plankton samples analyzed by Fontugne and Duplessy

(1981) indicates that is is difficult to define a representative mean value and an extreme value such as -17 $^o/_{oo}$ seems not impossible (Müller et al., 1983). On the other hand, common land plants ("C3 plants") have $\delta^{13}C_{org}$ values 5 to 10 $^o/_{oo}$ lower (Fig. 21). Difficulties in interpretation of the data may also occur because of changes in isotope composition during diagenetic alteration of the organic matter, temperature-dependence of carbon isotopic fractionation by phytoplankton, and changes in importance of C3 and C4 plants (cf., Fig. 21). Furthermore, the $\delta^{13}C_{org}$ value of any plant is dependant upon the $\delta^{13}C_{org}$ value of the CO_2 it uses for photosynthesis. Thus, changes in the CO_2 reservoir through geological time have resulted in different $\delta^{13}C_{org}$ values in the biomass (Welte et al., 1975; Waples, 1981; Arthur et al., 1985).

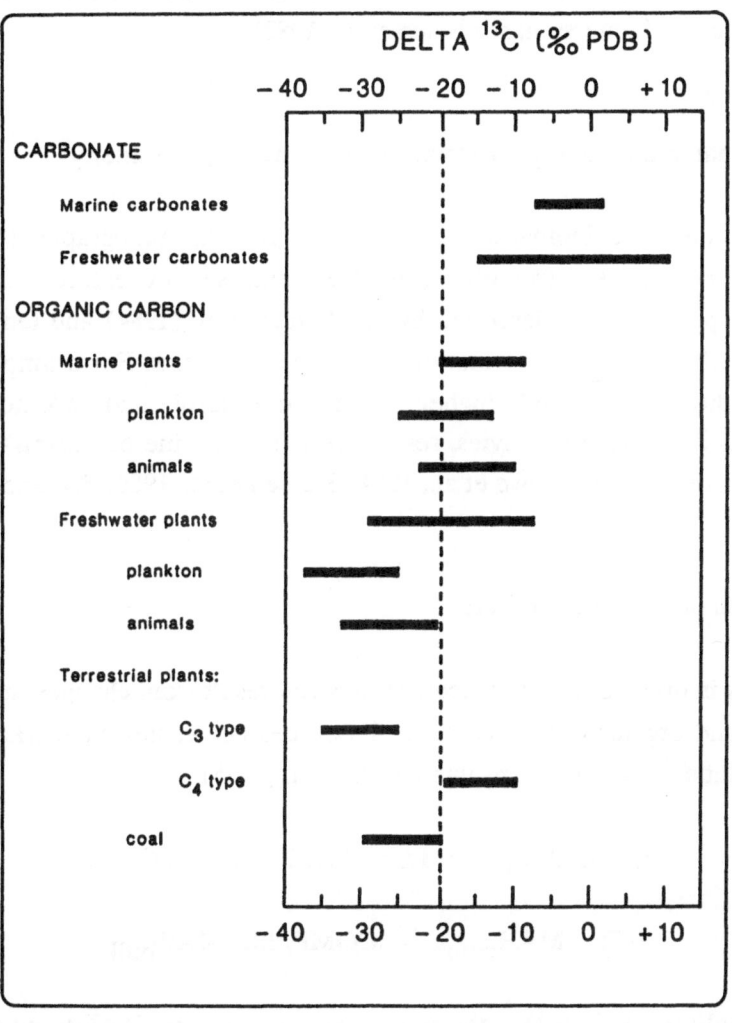

Fig. 21: $\delta^{13}C$ values of different types of carbonate and organic carbon (after Waples, 1981).

(4) Kerogen microscopy

Kerogen was concentrated by treating sediment samples successively with HCl and HF to remove carbonates and silicates, respectively (Durand, 1980). Maceral analyses were performed on embedded kerogen concentrates by a 2D-scanning method in reflected light and in fluorescent light (see Littke, 1987). In order to characterize the source of the organic matter, the macerals vitrinite, inertinite, sporinite, liptodetrinite, and alginite were determined (classification according to the nomenclature described by Hutton et al. (1989) and Stach et al. (1982)).

To assess maturity, vitrinite reflectivities were measured using particles larger than 10 μm (λ = 546 nm, oil immersion; see Stach et al., 1982).

(5) Gas chromatography and gas chromatography/mass spectrometry

In order to determine biomarker distributions, gas chromatography (GC) and gas chromatography/mass spectrometry (GC/MS) analyses of extracts from selected samples were performed as described by Rullkötter et al. (1984) and ten Haven et al. (1989). Biomarkers provide information on source organisms. For example, long-chain n-alkanes indicate terrestrial higher plants, dinosterol and alkenones indicate dinoflagellates and prymnesiophytes, respectively (i.e., marine organisms), as source of the organic matter (e.g., Marlowe et al., 1984; Brassell et al., 1986; Bruland et al., 1989).

(6) Calculation of accumulation rates

Since changes in organic carbon concentrations can result from changes in both mineral components and organic carbon content, the percentage values were transformed into mass accumulation rates, following van Andel et al. (1975):

$$(6) \quad MAR_{bulk} = LSR \ (WBD - 1.026 \ PO/100)$$

$$(7) \quad MAR_{comp} = COMP/100 \ MAR_{bulk}$$

where MARbulk (g cm^{-2} ky^{-1}) is the bulk accumulation rate, MARcomp (g cm^{-2} ky^{-1}) the mass accumulation rate of a single sediment component, LSR (cm ky^{-1}) the linear sedimentation rate, WBD (g cm^{-3}) the wet bulk density, PO (%) the porosity, and COMP (%) the amount of a single sediment component (e.g., organic carbon).

Using these accumulation rates, which are expressed in units of mass per area and time (i.e., gC cm^{-2} ky^{-1}), dilution effects by mineral components can be excluded.

4.2. Comparison of organic matter characteristics derived from different techniques

The use of a single method for determining the organic carbon composition may yield results which can be interpreted in different ways. Thus, in this study at least three of the methods summarized in Figure 22 were used to characterize the organic matter fraction.

COMPOSITION OF ORGANIC MATTER

*** ELEMENTAL ANALYSIS:
  C/N RATIO OF BULK SEDIMENT;
  H/C and O/C RATIOS OF KEROGEN CONCENTRATES

*** ROCK-EVAL PYROLYSIS:
  HYDROGEN INDEX vs. OXYGEN INDEX

*** KEROGEN MICROSCOPY

*** CARBON STABLE ISOTOPES
  OF ORGANIC CARBON

*** GAS CHROMATOGRAPHY (GC)

*** GASCHROMATOGRAPHY /
  MASS SPECTROMETRY (GC/MS)

Fig. 22: Summary of methods used for determination of organic carbon composition.

In organic-carbon-rich sediments, such as black shales, the amount of marine/terrigenous organic carbon can be roughly estimated from hydrogen index values, as shown, for example, for DSDP Atlantic Ocean black shales (Fig. 23; Stein et al., 1986). The recent data from ODP-Sites 645 and 658 also fit into this correlation (Fig. 23). When interpreting hydrogen index values in terms of organic matter origin it has to be considered, however, that low hydrogen index values particularly for organic-matter-lean sediments may also result from mineral matrix effects (Katz, 1983; Espitalié et al., 1984). Rock-Eval pyrolysis performed on kerogen concentrates from the same samples may yield distinctly higher hydrogen index values (Fig. 24). Thus, before interpreting low hydrogen index values in terms of terrigenous organic carbon dominantly being present in the organic fraction, one should use additional methods to support this interpretation, such as maceral composition, $\delta^{13}C_{org}$, or C/N ratios. Also, there is no simple correlation between hydrogen index values and C/N ratios (Fig. 25). This is not surprising because C/N ratios of terrigenous organic matter display broad variations (cf., Fig. 19) making an absolute estimate of marine/terrigenous organic matter very difficult.

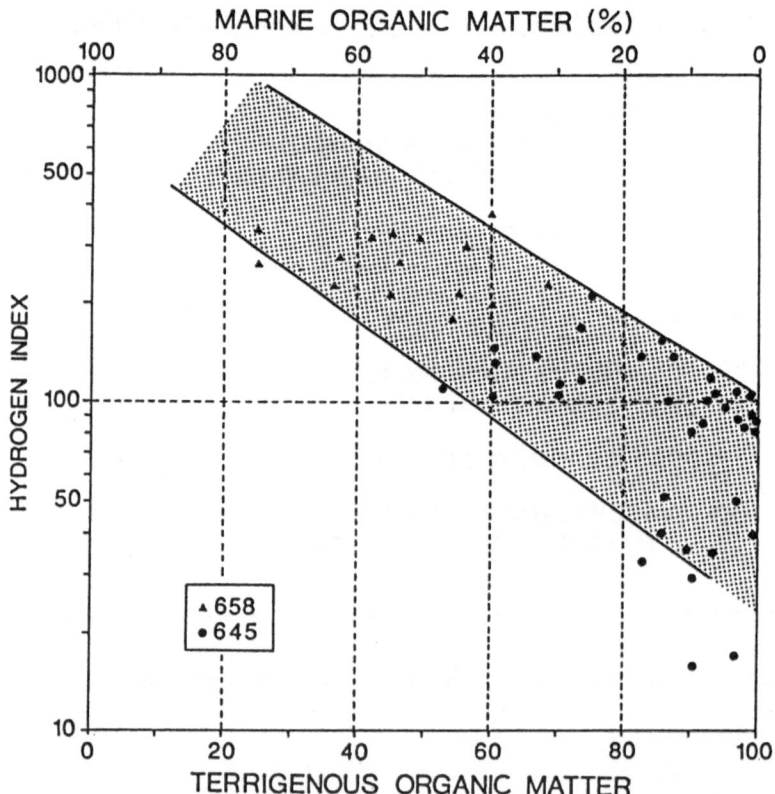

Fig. 23: Correlation between hydrogen index derived from Rock-Eval pyrolysis and maceral data (stippled field), based on DSDP data (Stein et al., 1986). Data from Sites 645 and 658 are plotted into the correlation lines.

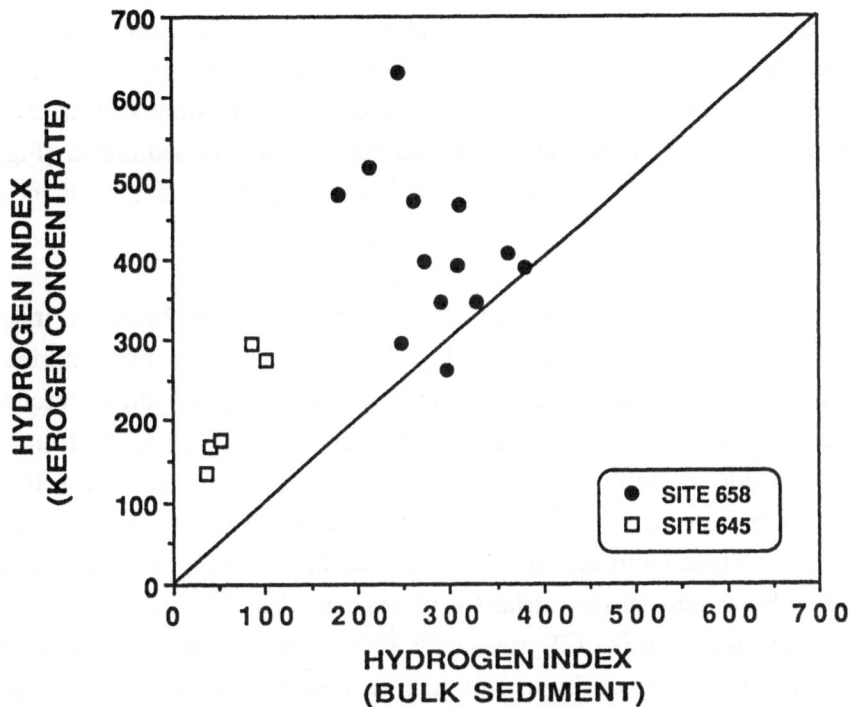

Fig. 24: Correlation between hydrogen indices measured on bulk sediment samples and hydrogen indices (mgHC/gC) measured on kerogen concentrates.

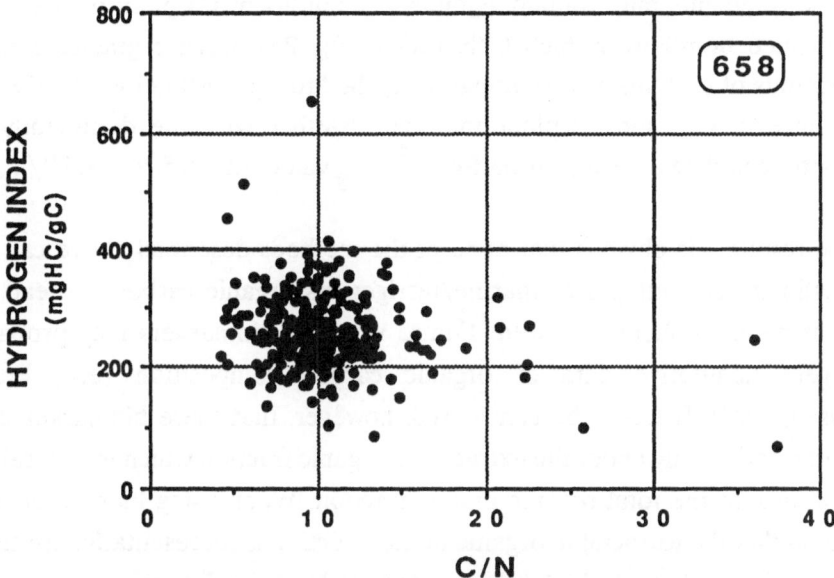

Fig. 25: Correlation between hydrogen indices (mgHCgC) measured on bulk sediment samples and C/N ratios (Site 658).

In order to compare results of different techniques in some more detail various methods described in Chapter 4.1. were applied to a selected set of samples from ODP-Site 658 (cf., Tab. 6). In general, the results of all methods are consistent in that they clearly indicate the dominance of marine organic carbon in the 658 sediments (Fig. 26; cf., Chapter 6.4.2., Figs. 76 to 78). However, differences occur when details of the data are looked at.

While $\delta^{13}C_{org}$ values of about -20.5 to -19.5 $^o/_{oo}$ (Fig. 26) and hydrogen index values determined on kerogen concentrates (cf. Fig. 76) point to almost pure marine organic carbon in most of the samples, maceral composition, hydrogen indices determined on bulk sediments, and C/N ratios suggest the presence of significant amounts of terrigenous organic carbon (Figs. 23 and 26). The latter is also corroborated by biomarker composition (cf., Fig. 81; ten Haven et al., 1989). When interpreting $\delta^{13}C_{org}$ values, it has to be taken into account that relatively high $\delta^{13}C_{org}$ values of about -18 to -10 $^o/_{oo}$ may be caused by terrestrial C4 plants (cf., Fig. 21). Although most modern terrestrial plants belong to the C3 group with $\delta^{13}C_{org}$ values in the range of -34 to -24 $^o/_{oo}$ (Fig. 21), in extreme environments such as deserts and swamps, C4 plants (e.g., grass) may become important occasionally (Smith, 1976; Waples, 1981). Because the source area of the terrigenous organic matter deposited at Site 658 was the Saharan desert, these sediments may contain significant amounts of C4-plant debris. This may also be reflected by relatively high C/N ratios (Fig. 26) because grasses contain low relative amounts of proteins when compared to plankton (cf., Müller et al., 1983). Thus, the occurrence of C3- and C4-plant material together with the dominantly marine organic matter could be an explanation for $\delta^{13}C_{org}$ values of -20.5 to -19.5 $^o/_{oo}$, too.

From these results it is obvious that none of the methods described above can be used easily to estimate percentages of marine/terrigenous organic carbon. Instead, several independant methods should be used. Future work on biomarkers may probably also help to get quantitative data of organic carbon composition (e.g., Prahl and Muehlhausen, 1989). It has to be considered, however, that these biomarker data give detailed information only about the extractable organic fraction which is generally only a small proportion of the total organic matter fraction. When using biomarker data one has to assume that the extractable organic matter fraction is representative for the entire organic matter in a sediment which is known to not always be the case.

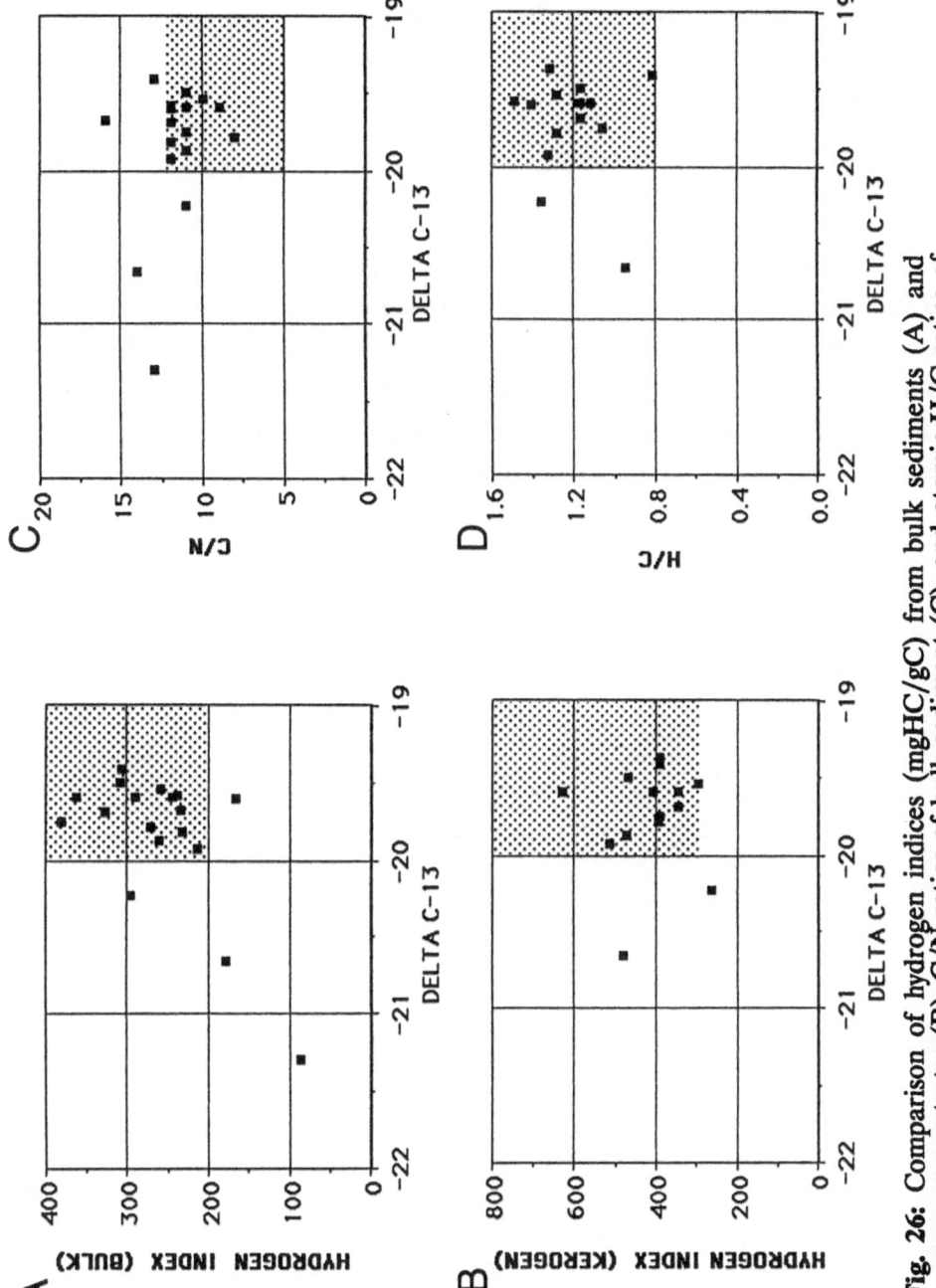

Fig. 26: Comparison of hydrogen indices (mgHC/gC) from bulk sediments (A) and kerogen concentrates (B), C/N ratios of bulk sediment (C), and atomic H/C ratios of kerogen concentrates (D) with $\delta^{13}C_{org}$ values ($^0/_{00}$ PDB). Data produced on the same samples from Site 658. Stippled area marks "marine organic matter". ($^{13}C_{org}$ measurements were performed by P. Müller, Bremen University)

5. Accumulation of organic carbon in Baffin Bay and Labrador Sea sediments (ODP-Leg 105)

5.1. Introduction

During ODP-Leg 105, holes were drilled at three sites in Baffin Bay and the Labrador Sea (Fig. 27). Major objectives of this drilling campaign were: (1) to study the tectonic development of these basins, (2) to study the evolution of surface- and deep-water circulation in these basins and their connection to the Arctic and Atlantic Oceans, and (3) to study the record of high-latitude paleoclimate and the timing and frequency of climatic changes, particularly of glacial/interglacial cycles (Srivastava, Arthur et al., 1987).

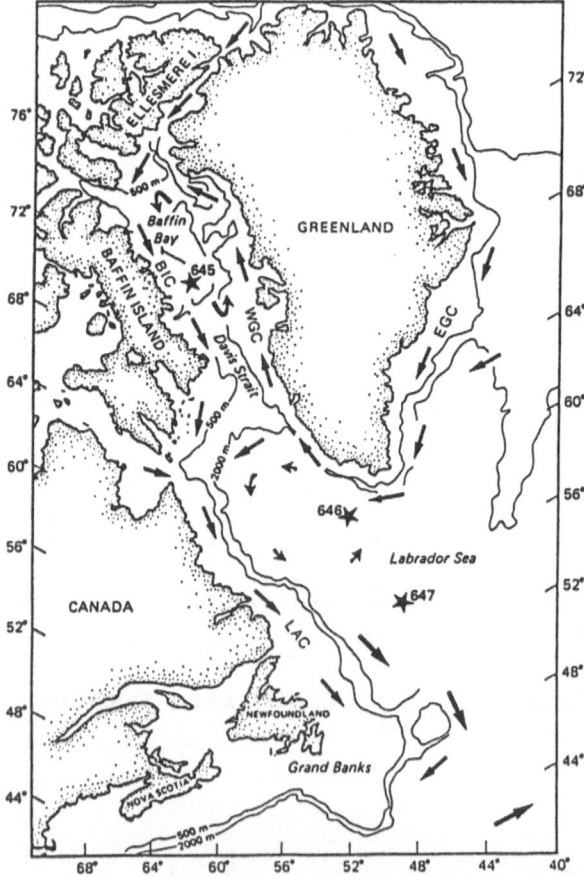

Fig. 27: Location map of Sites 645, 646, and 647. Black arrows indicate surface-water circulation (according to Tchernia, 1982). BIC = Baffin Island Current; LAC = Labrador Current; EGC = East Greenland Current; WGC = West Greenland Current.

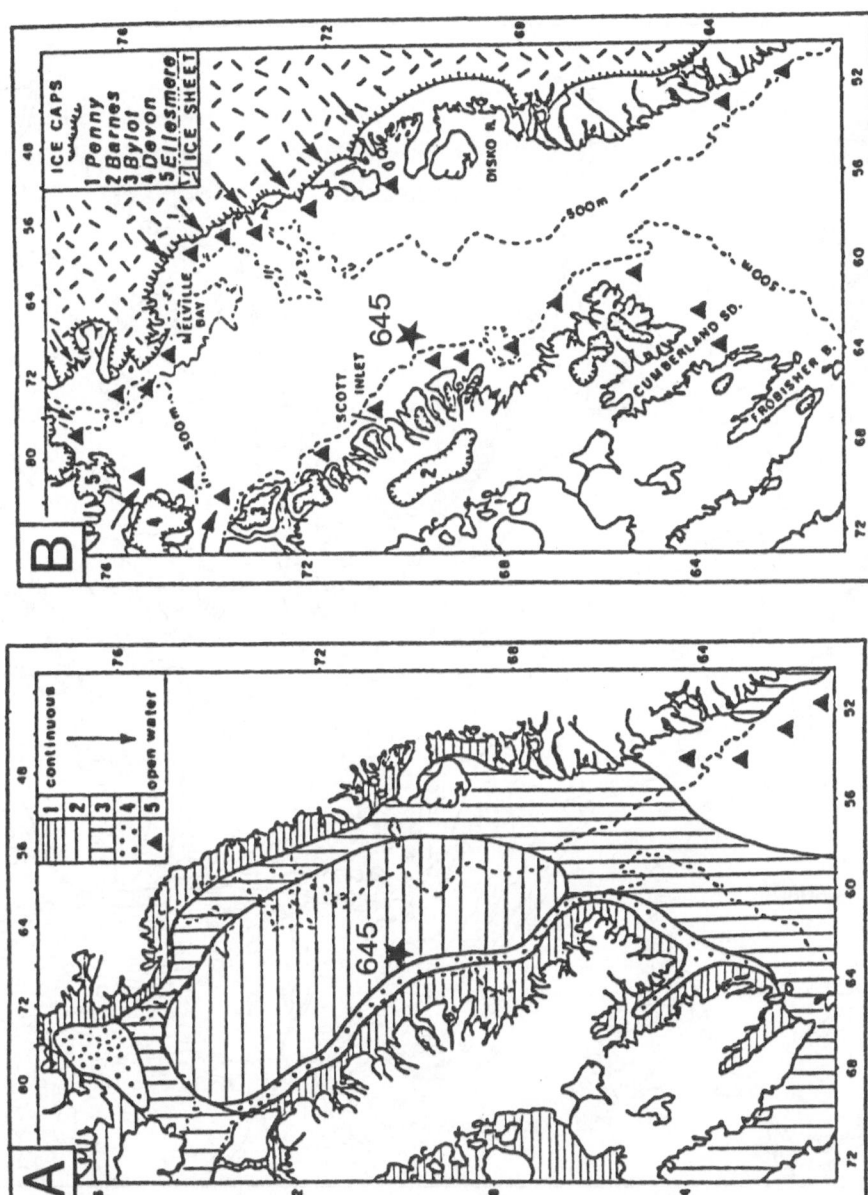

Fig. 28: (A) Distribution of sea ice in Baffin Bay during winter (after Osterman, 1980). (1) continuous ice cover; (2) consolidated pack ice; (3) unconsolidated pack ice; (4) semi-permanent open water; and (5) open water with icebergs. (B) Major iceberg concentration during summer. Arrows show main sources of ice bergs (after Pelletier et al., 1975). (from Mudie and Short, 1985).

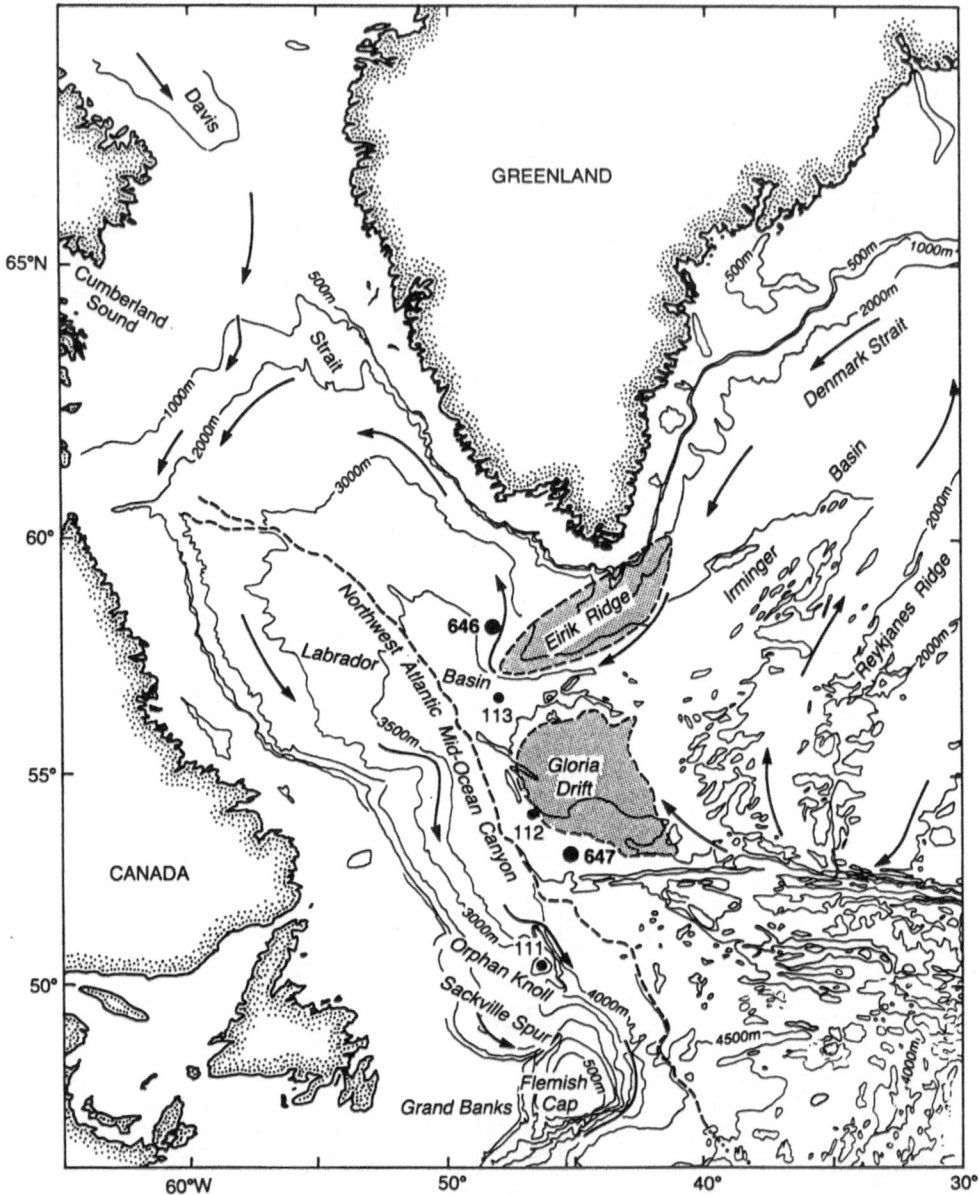

Fig. 29: Deep-water circulation in the Labrador Sea with location of Sites 646 and 647. Erik Ridge and Gloria Drift = major sediment drifts (Srivastava, Arthur, et al., 1987).

Baffin Bay and the Labrador Sea probably were the main connection between the North Atlantic and the Arctic Ocean before the opening of the Greenland and Norwegian Seas (Srivastava et al., 1981). Thus, the distinct paleoenvironmental changes that occurred in this region during Tertiary times (such as the tectonic development of new oceanic basins and the development of glacial climate in the Arctic area) should have resulted in distinct changes in sediment composition. Detailed sedimentological, paleontological, and geochemical studies of deep-sea sediments from Baffin Bay and the Labrador Sea (as well as from the Norwegian Sea; see Leg 104 results in Eldholm, Thiede, et al., 1989), therefore, provide a unique possibility for reconstructing paleovironmental changes in the North Atlantic/Arctic Ocean region.

The data presented here mainly concentrate on the changes in quantity and composition of organic matter with time. Emphasis was placed on the relationships between organic carbon accumulation and both paleoclimatic and paleoceanic conditions in Baffin Bay and Labrador Sea through late Cenozoic times.

5.2. Recent and sub-recent environments in Baffin Bay and Labrador Sea

Baffin Bay, located between Greenland and Baffin Island (Fig. 27), is connected to the Arctic Ocean in the North by the narrow Nares Strait and to the Atlantic Ocean in the South by the shallow Davis Strait (less than 600m water depth). The Nares Strait has probably been opened during Middle Miocene times allowing the exchange of water masses between the Arctic Ocean and Baffin Bay since that time (Srivastava et al., 1981). In the center of Baffin Bay, water depths exceed 2300m. Surface-water circulation in Baffin Bay is dominated by currents entering the bay from the Arctic Ocean (the Baffin Island Current) and the North Atlantic Ocean (the West Greenland Current) (Fig. 27; Tchernia, 1982; Williams, 1986). This circulation is counter-clockwise (cyclonic) (Fig. 27; Marko et al., 1982; Williams, 1986). Between December and April, Baffin Bay is covered by sea ice, whereas during the rest of the year, drifting icebergs are of major importance (Fig. 28; Pelletier et al., 1975; Osterman, 1982; Marko et al., 1982; Williams, 1986).

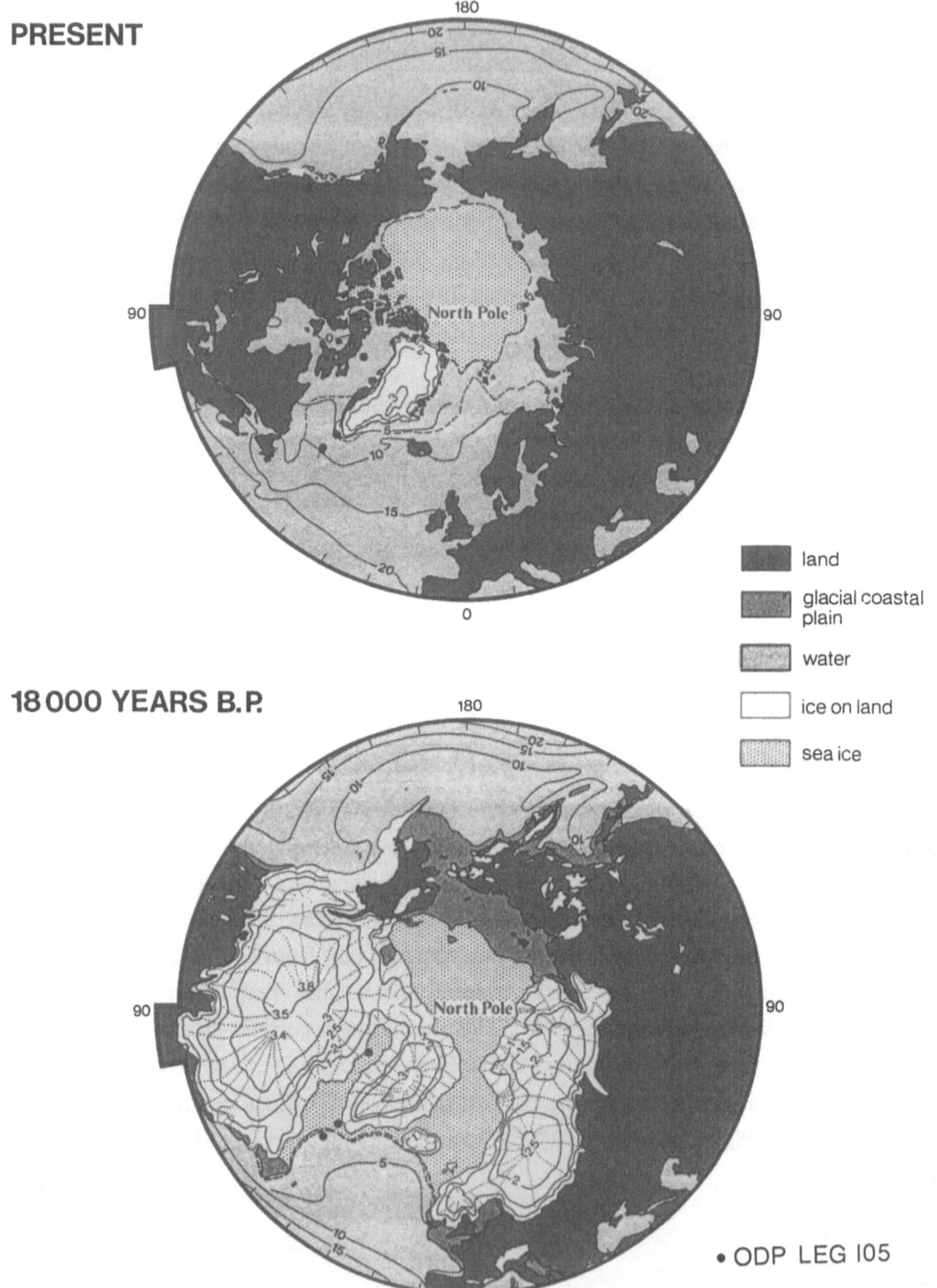

PRESENT

18 000 YEARS B.P.

land

glacial coastal plain

water

ice on land

sea ice

• ODP LEG 105

Fig. 30: The ice situation of the Northern Hemishpere for the Present and for the Last Glacial Maximum (18000 yrs.BP). (from Degens, 1989 after CLIMAP, 1976, 1981).

In the Labrador Sea, a deep basin in the Northwest Atlantic Ocean with water depths of more than 3500 m, surface-water circulation is dominated by the Greenland Current System (i.e., the East and West Greenland Currents; Fig. 27). The Labrador Sea is a zone of active mixing between polar waters and waters of tropical origin. Sea ice and iceberg occasionally occur in winter and spring (Tchernia, 1982). The deep-water circulation in the Labrador Sea is mainly influenced by the Denmark-Strait Overflow Water (DSOW) and the Iceland-Scotland Overflow Water (ISOW) which presently spill over the Greenland-Iceland-Faeroe Ridge and which forms a major component of the North Atlantic Deep Water (NADW) (Fig. 29). This southward flowing overflow water (as contour current) follows the complex morphology of the ocean floor and is responsible for the formation of major sedimentary bodies ("sediment drifts") (Fig. 29; McCave and Tucholke, 1986). To study the history of these deep-current influenced sediment drifts was also one of the major objectives at Sites 646 and 647 (Srivastava, Arthur et al., 1987).

Compared to the modern situation described above, the climatic and oceanic conditions were very different 18000 years ago, i.e., during the last glacial maximum (Fig. 30). Based on the landmark study of the CLIMAP group (CLIMAP, 1976; 1981), detailed information is available about these differences. It is well known that the world of the last glacial maximum was characterized by (1) increased extent and volume of continental ice, (2) lowered sea level and reduced oceanic surface area, (3) increased sea-ice cover in high latitudes, (4) increased surface albedo, and (5) reduced sea-surface temperature. These differences between the modern and the last glacial maximum conditions are greatest in high latitudes, moderate in equatorial and boundary current regions, and minimal in the central gyres (CLIMAP, 1981). Thus, the high-latitude area of Leg 105 is a very sensitive area for paleoenvironmental changes and, therefore, optimal for studying short- as well as long-term changes in paleoclimate and paleoceanic circulation.

Tab. 3: Geographic locations and water depths of ODP-Leg 105 Sites

ODP-Site	Location	Water depth
645	70°27.48'N, 64°39.30'W	2018 m
646	58°12.56'N, 48°22.15'W	3451 m
647	53°19.88'N, 45°15.72'W	3862 m

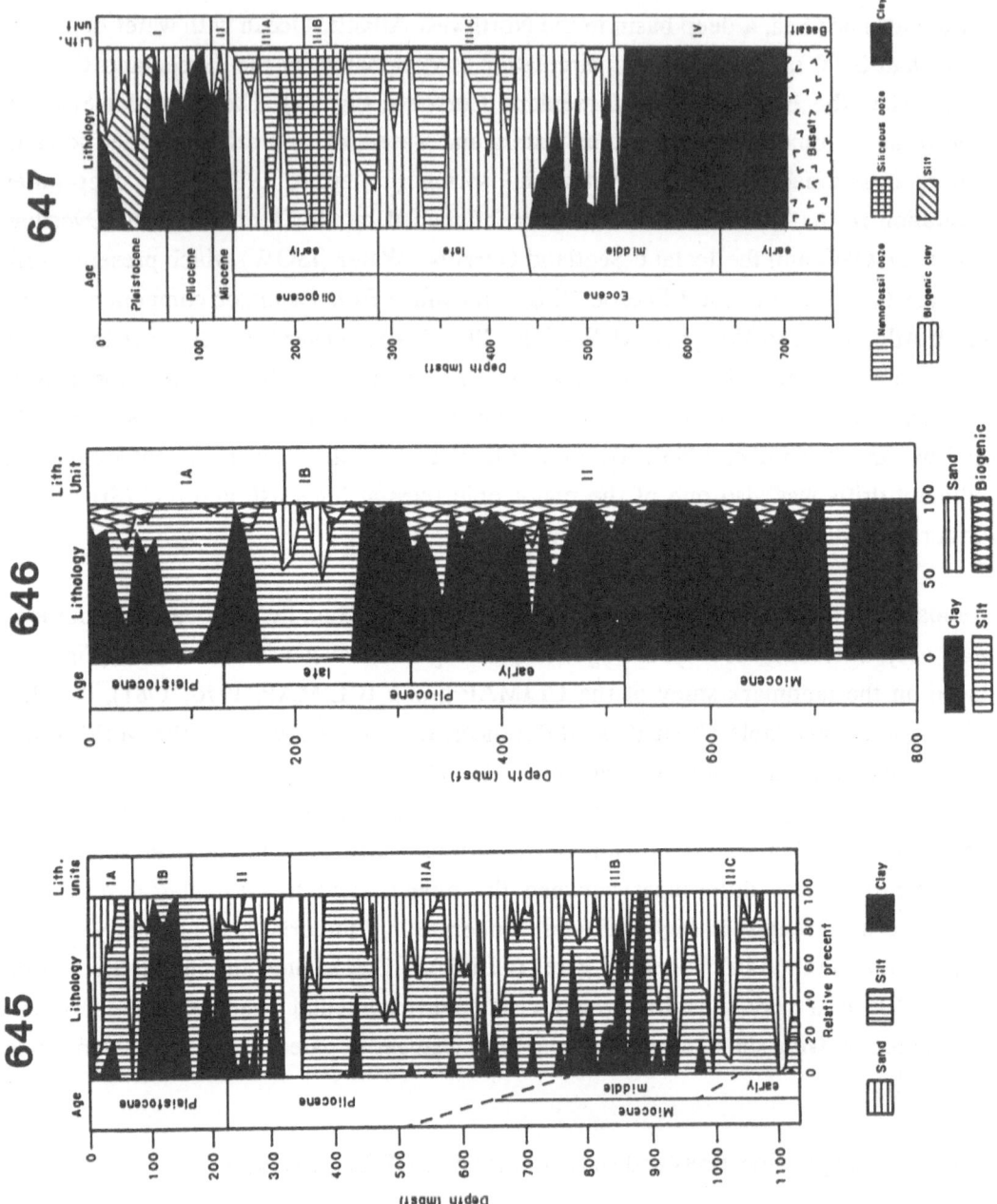

Fig. 31: Summary diagrams of lithologic units at Sites 645, 646, and 647 (Srivastava, Arthur et al., 1987).

5.3. Geological setting and sediments

Drilling at Site 645 was performed on the western slope of Baffin Bay (Fig. 27, Tab. 3). Total depth of penetration was 1147.1 meters below sea floor (mbsf). Sediments are of early Miocene to Holocene age and predominantly consists of terrigenous clay, silt and sand; biogenic components are only of minor importance (Fig. 31). Thus, the biostratigraphic time control is less precise than that defined at Sites 646 and 647 (Baldauf et al., 1989). Abundant dropstones characterize the upper 465 m (i.e., the last 3.4 Ma); sporadic ice rafting can be dated back to about 8.5 Ma (Arthur et al., 1989; Korstgärd and Nielsen, 1989). Furthermore, the sediments (for normal open-marine sediments) have unusually high organic carbon contents. The sediment sequence at Site 645 is characterized by high mean sedimentation rates of 28 to 135 m/my (Fig. 32).

Site 646 was occupied on the northern flank of Eirik Ridge, Labrador Sea (Figs. 27 and 29; Tab. 3). Total depth of penetration was 766.7 mbsf. Sediments are of late Miocene to Holocene age (Fig. 31). The lower part of the sequence (766.7 to 236.1 mbsf) mainly consists of greenish-gray clay (claystone) and silt (siltstone), whereas the upper 236.1 m (i.e., the last 2.5 Ma) are characterized by dark gray to light gray clays and silts with carbonate contents of up to 40% and a common occurrence of dropstones (Srivastava, Arthur et al., 1987; Korstgärd and Nielsen, 1989). Mean sedimentation rates are high and vary between 49 and 96 m/my (Fig. 32).

At Site 647 in the southern Labrador Sea south of the Gloria Drift (Figs. 27 and 29; Tab. 3), a 699 m thick sediment sequence of early Eocene to Holocene was penetrated. At 699 mbsf, the basaltic basement was reached (Srivastava, Arthur et al., 1987). The sediments are dominated by terrigenous material in the upper 116 mbsf and again below 530 mbsf. The middle part of the sequence is characterized by biogenic carbonate and siliceous oozes with a maximum occurrence of biogenic silica between 250 and 150 mbsf (Fig. 31; Bohrmann and Stein, 1989). Abundant dropstones occur in the upper 116 m (i.e., the last 2.5 Ma) (Korstgärd and Nielsen, 1989; Stax and Stein, 1989). Mean sedimentation rates vary between 16 and 45 m/my. Major hiatuses and/or extremely low sedimentation rates occur in the Miocene interval (Fig. 32).

Both Labrador-Sea Sites 646 and 647 were chosen to avoid thick turbidite sequences that characterize the region of the Northwest Atlantic Mid-Ocean Channel (NAMOC) (Fig. 29; Srivastava, Arthur et al., 1987). Thus, turbidites only occasionally occur in the sediment sequences of Sites 646 and 647.

These turbidites can be identified as thin laminated detricarbonate beds and are interpreted as spillover facies from the NAMOC indicating pulses of major NAMOC activities (Chough and Hesse, 1987; Srivastava, Arthur et al., 1987). Because of their mineralogical and structural characteristics, the turbidites can be easily distinguished from the normal pelagic sedimentation. This is important for high-resolution studies of paleoclimate and paleoceanography.

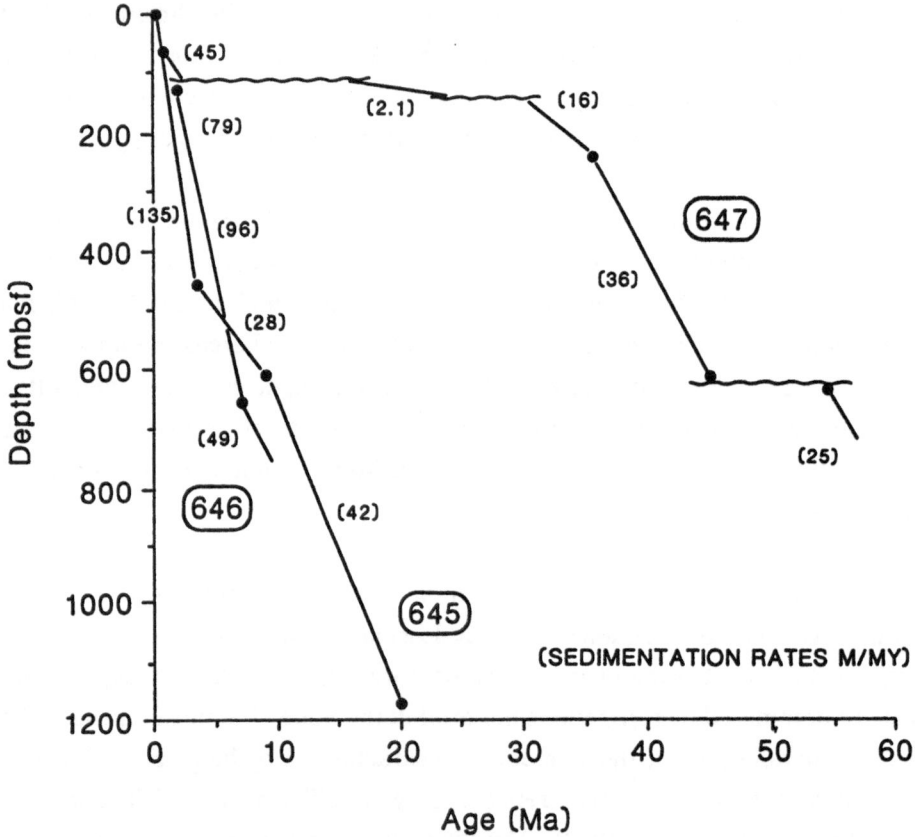

Fig. 32: Sedimentation rates at Sites 645, 646, and 647 (after Baldauf et al., 1989).

5.4. Results

5.4.1. Quantity of organic carbon and carbonate content

Almost all of the sediment sequence at Site 645 is characterized by TOC values that are distinctly higher than those recorded in normal open-marine environments. Most of the TOC values exceed 0.5% and range between 0.5% and 3% (Fig. 33). The lower values are concentrated in the upper 320 m, i.e., in the upper Pliocene to Holocene sediments. Between 780 and 340 mbsf (late Miocene to early Pliocene), most TOC contents vary between 0.5 and 1.5%. Maximum organic carbon contents occur between 780 and 1040 mbsf in middle Miocene sediments with values ranging between 1 and 3% . In the lowermost 100m of the sequence, TOC values are lower again (0.5 to 1.2%; Fig. 33).

Although the upper part of the sequence characterized by lower TOC values, displays higher carbonate contents (Fig. 33), no straightforward negative correlation between organic carbon and carbonate content is obvious (Figs. 34 and 35). Especially in the Miocene interval organic carbon contents show high-amplitude variations between 0.5 and 3% whereas the carbonate contents are constantly low (less than 5%; Fig. 35).

In order to be able to look at short-term fluctuations in organic carbon (and carbonate) contents, some intervals were sampled and studied in detail. Based on mean sedimentation rates (Baldauf et al., 1989), these fluctuations have periods of 20000 to 100000 years (Stein et al., 1989a; Fig. 35).

The open-ocean Sites 646 and 647 are characterized by predominantly low TOC values ranging between less than 0.1 and 0.75 % (Figs. 36 and 37). At Site 646, low-amplitude variations in organic carbon ranging between 0.25 and 0.45% occur in the upper Miocene to lower Pliocene section, with the lower values in the lower part of the section (Fig. 36). The following upper Pliocene to Holocene interval displays high-amplitude variations between 0.15 and 0.75 % TOC. In the right half of Figure 36, short-term fluctuations are shown for three intervals in detail, also suggesting the presence of periods of 20000 to 100000 years.

At Site 647, low TOC values of 0.05 to 0.3 % are dominant in the Eocene time interval (Fig. 37). Some higher organic carbon contents (0.3 to 0.55 %) are concentrated in the early Oligocene and late Pliocene. In the Miocene and Pleistocene time intervals, the TOC values again are low (less than 0.3 %; Fig. 37).

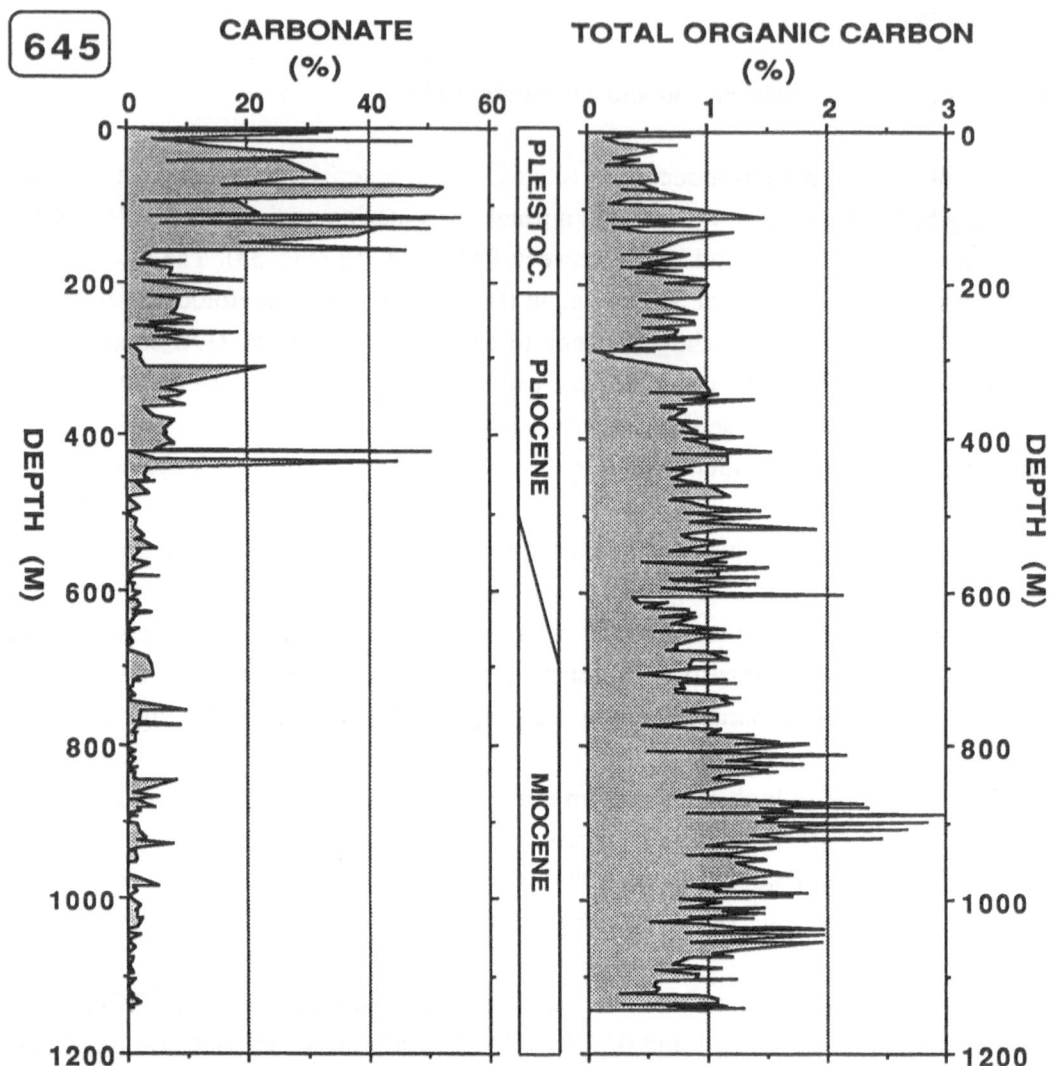

Fig. 33: Carbonate and total organic carbon contents at Site 645.

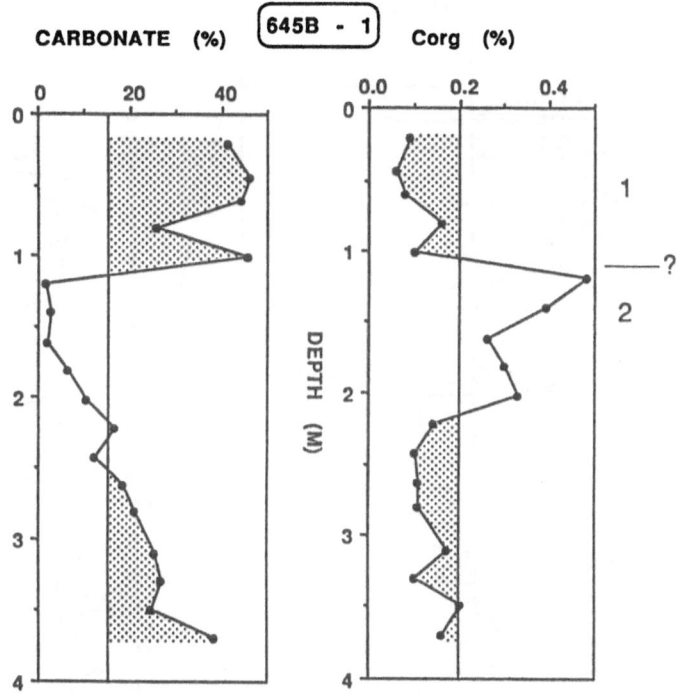

Fig. 34: Carbonate and total organic carbon contents of the upper 4 m of Site 645B. 1/2 = assumed oxygen isotope stage boundary.

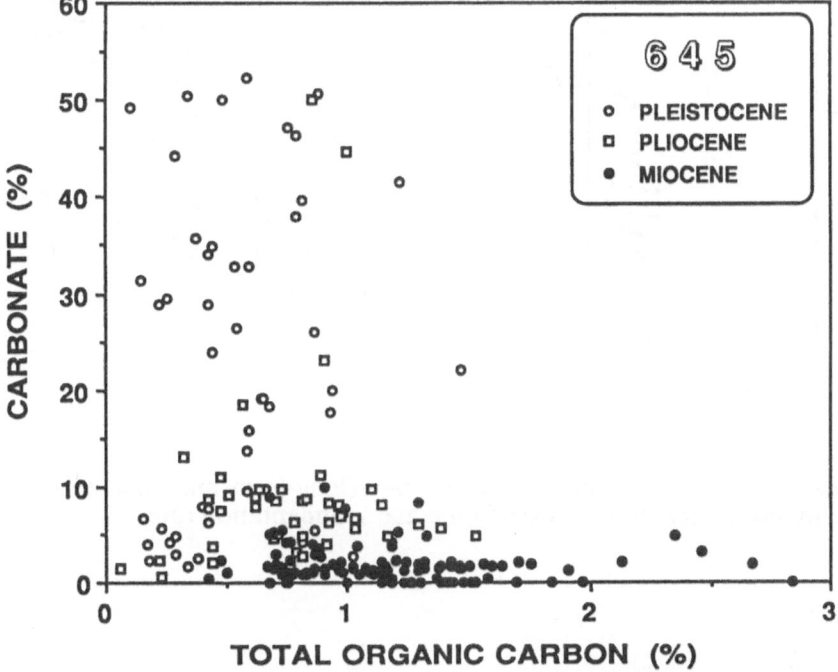

Fig. 35: Correlation between carbonate and total organic carbon contents at Site 645.

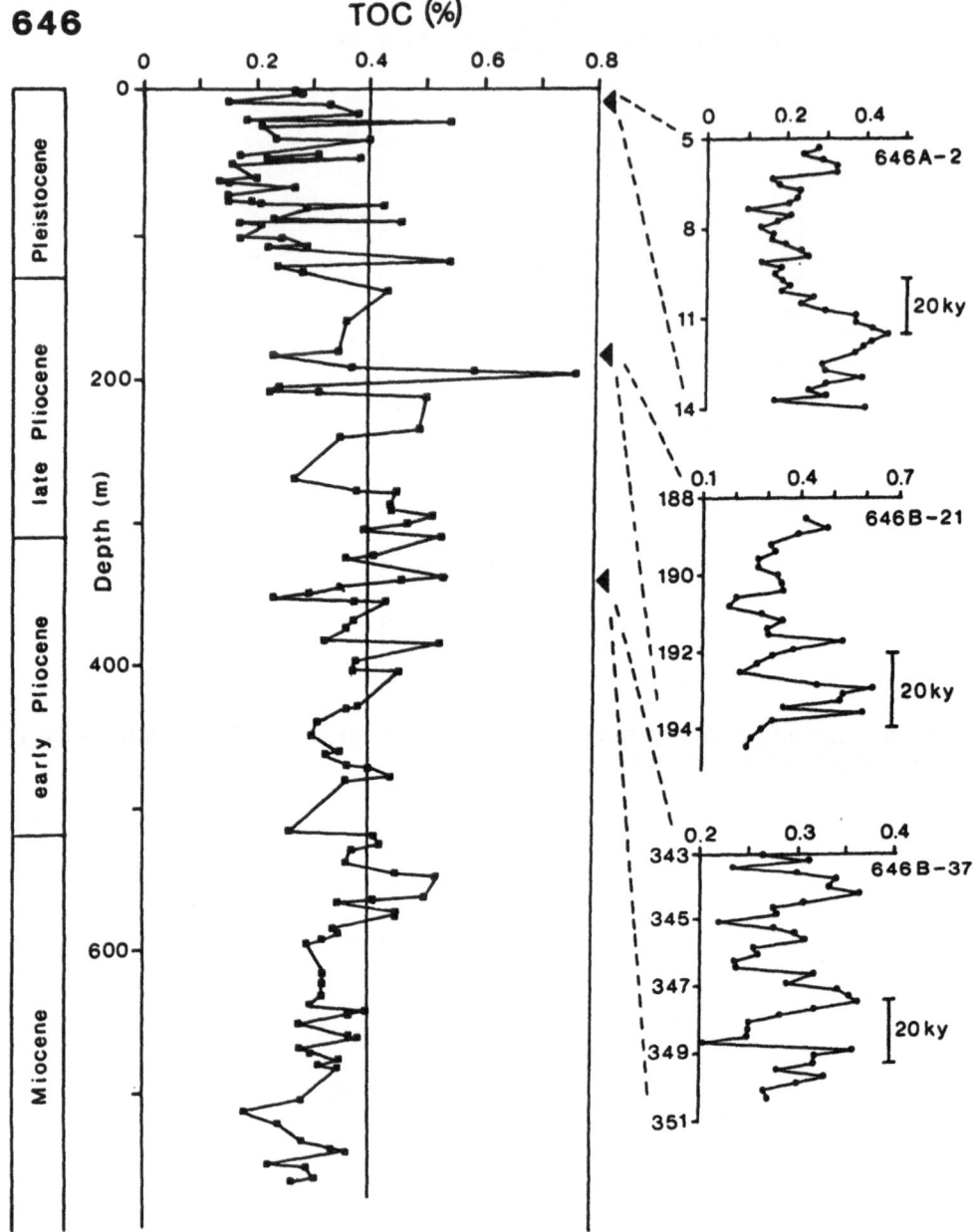

Fig. 36: Total organic carbon content at Site 646 (long-term and short-term records). Vertical bar marks a 20ky interval, based on mean sedimentation rate.

53

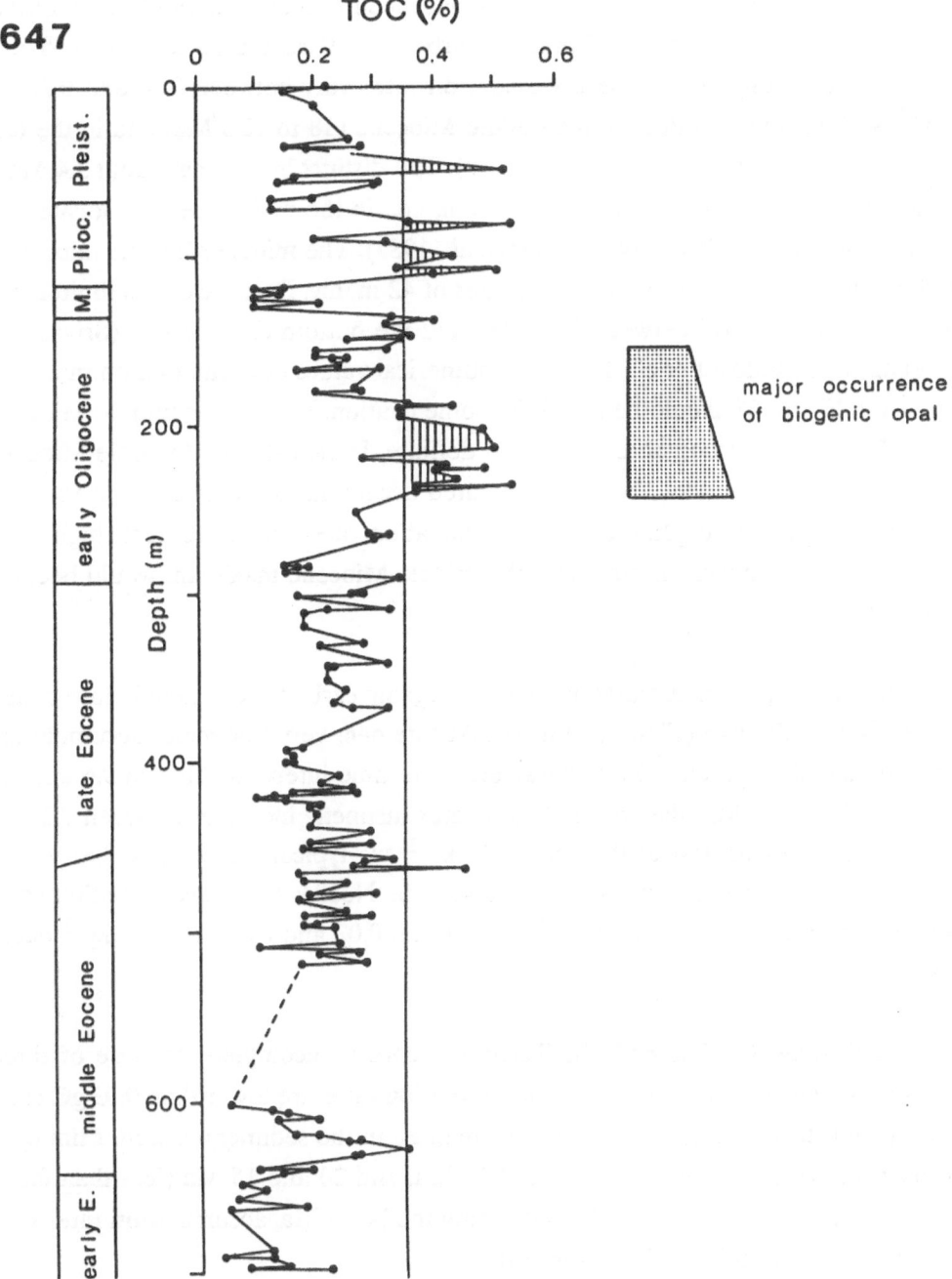

Fig. 37: Total organic carbon content at Site 647. Stippled field marks interval of major occurrence of biogenic opal according to Bohrmann and Stein (1989).

For intervals where bulk accumulation rates had been determined (Stein and Littke, 1990), mean accumulation rates of organic carbon were calculated for Site 645 and plotted versus age (Fig. 38). These accumulation rates vary between 0.04 and 0.20 gC cm^{-2} ky^{-1}, with maximum values in the middle Miocene (18 to 12.5 Ma) and in the late Pliocene to Holocene (the last 3.4 Ma) (Fig. 38). The distinct increase at about 3.4 Ma is also observed in the inorganic sediment fraction, i.e., in the mainly detricarbonate and siliciclastic components (Srivastava, Arthur et al., 1987). The middle Miocene maximum is based on minimum mean sedimentation rates of 42 m/my. Because of the limited age constraints for the interval between 750 mbsf and the bottom of Hole 645 (Srivastava, Arthur et al., 1987; Baldauf et al., 1989), the numerical values of accumulation rates and ages shown in Figure 38 should be used with some caution. However, a middle Miocene maximum for organic carbon accumulation is definite. If the estimated middle Miocene sedimentation rate is too low, then the estimated accumulation rates are also too low. This means that the real organic carbon accumulation rates may have been higher for the middle Miocene interval; in this case, the middle Miocene maximum would become even more distinct.

At Sites 646 and 647, the accumulation rates of organic carbon are distinctly lower than those recorded at Site 645 (Figs. 38 and 39). At Site 646, very low mean accumulation rates of less than 0.04 gC cm^{-2} ky^{-1} characterize the time intervals between 9.6 and 7.2 Ma (Fig. 39). Near 7.2 Ma, the accumulation rates distinctly increase. Between 7.2 and about 2.4 Ma, values of 0.04 to 0.10 gC cm^{-2} ky^{-1} are typical, with the higher values concentrating in the interval between 6.4 and 4.3 Ma. Near 2.4 Ma, accumulation rates of organic carbon decrease; values between less than 0.02 and 0.05 gC cm^{-2} ky^{-1} occur (Fig. 39).

At central Labrador Sea Site 647, the Tertiary record is incomplete because of three major hiatuses (Fig. 39). Organic carbon accumulation rates are lower than 0.03 gC cm^{-2} ky^{-1} throughout the entire time intervals documented in the sediment record. Minimum values occur between 55 and 45 Ma, 35 and 30 Ma, and 20 and 15 Ma (less than 0.003 gC cm^{-2} ky^{-1}). Between 45 and 35 Ma and during the last 3 Ma, accumulation rates vary between 0.01 and 0.03 gC cm^{-2} ky^{-1} (Fig. 39).

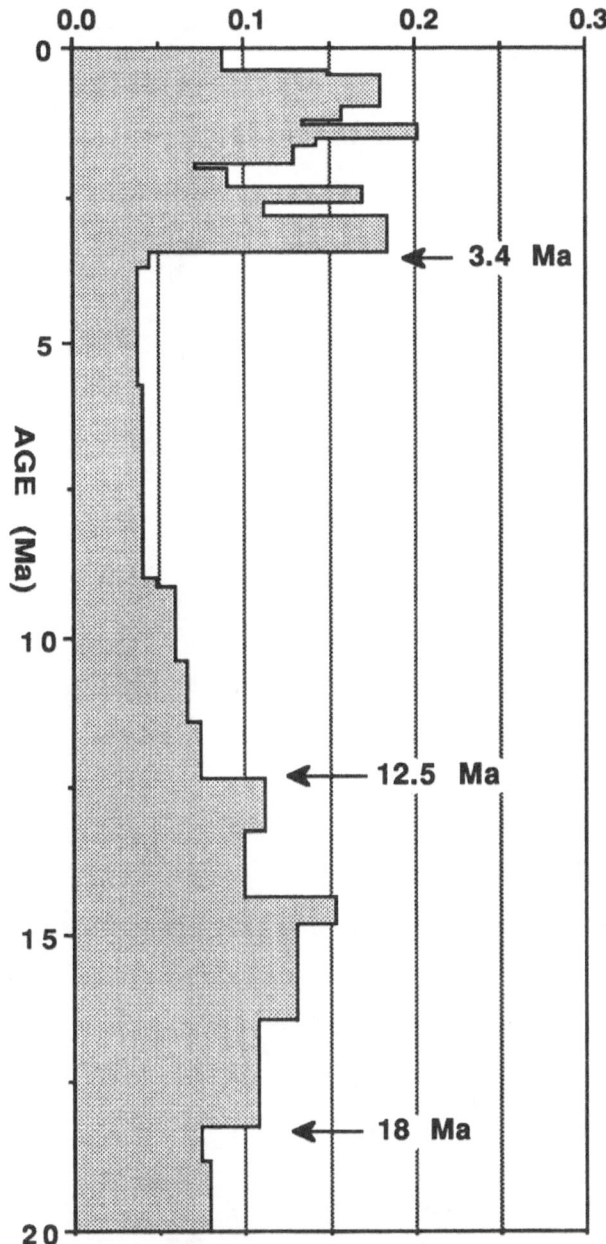

645 **ACCUMULATION RATES OF TOTAL ORGANIC CARBON (gC/cm2/1000yrs)**

Fig. 38: Accumulation rates of total organic carbon at Site 645.

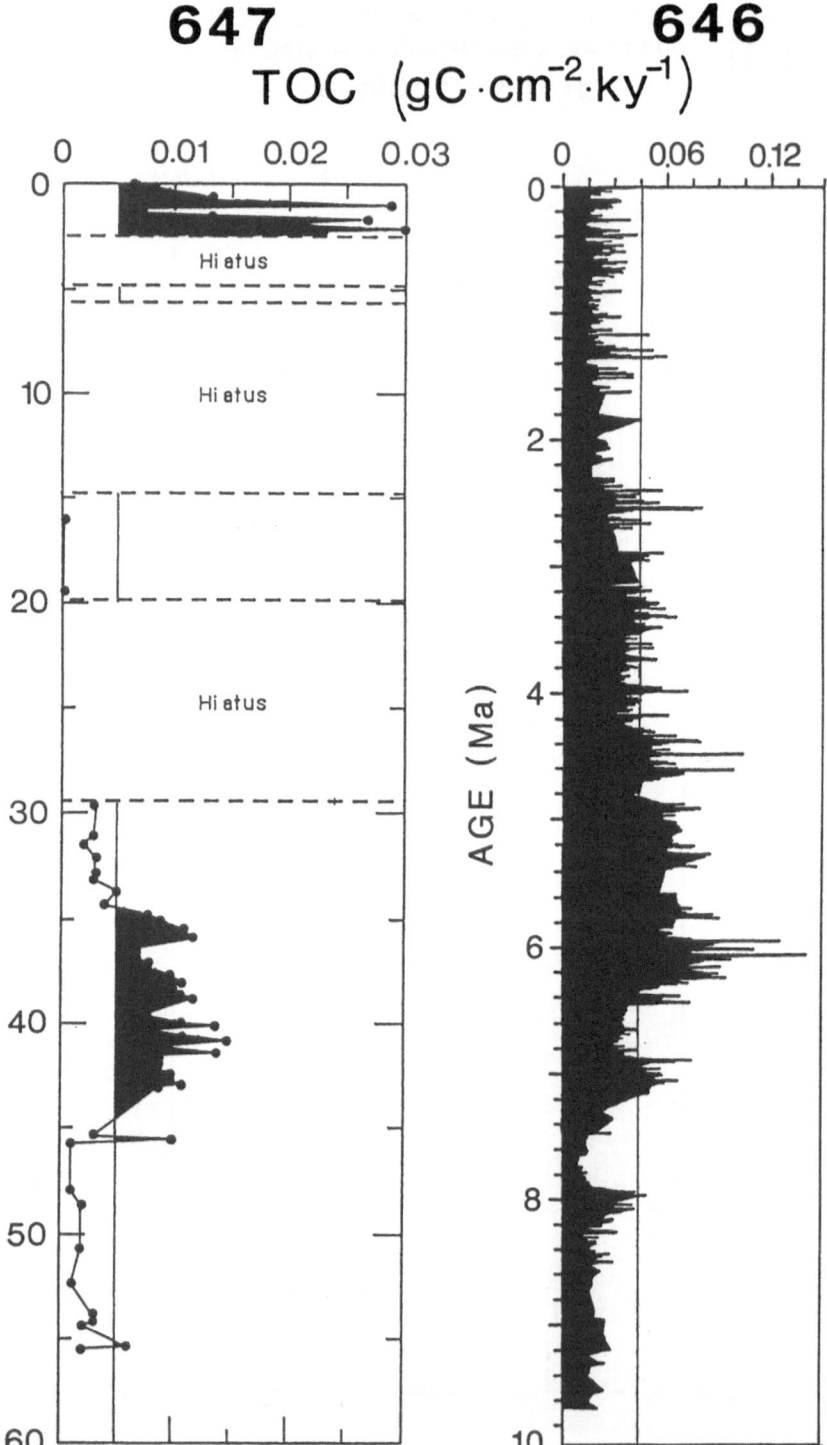

Fig. 39: Accumulation rates of total organic carbon at Sites 646 and 647. Record of Site 646 from Wolf and Thiede (1990) including data from Bohrmann (1988) and Stein et al., 1989a.

5.4.2. Composition of organic matter

Detailed studies on organic carbon composition were only performed on the sediment samples from Site 645; most of the samples from Sites 646 and 647 have organic carbon contents too low for detailed organic geochemistry investigations. As outlined in Chapter 4.2, it is important to use different independant methods to get information about the composition of the organic matter. Parameters used for determining the type of organic matter of Leg 105 sediments are organic carbon/nitrogen (C/N) ratios, hydrogen and oxygen index values, maceral composition determined by kerogen microscopy, and GC and GC/MS data (cf. Chapter 4 for description and comparison of methods).

Organic carbon/nitrogen ratios

The sediments of Site 645 are characterized by relatively high C/N ratios of generally more than 10. Most of the values vary between 10 and 30 (Fig. 40), indicating the predominance of terrigenous organic matter at Site 645. However, values around 10 may imply some admixture of marine compounds.

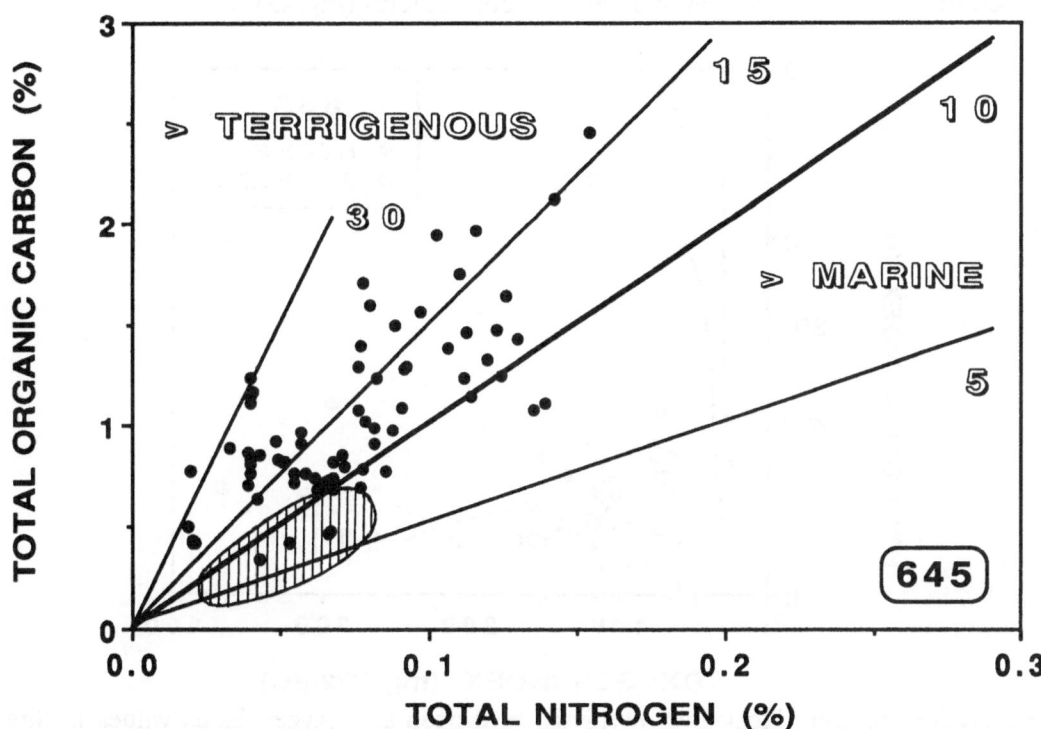

Fig. 40: Total organic carbon vs. total nitrogen contents at Site 645. Hatched areas mark data from Labrador-Sea Sites 646 and 647. Lines indicate C/N ratios (30, 15, 10, and 5).

58

At Sites 646 and 647, the C/N ratios varying between 10 and 4 (Fig. 40) are distinctly lower than those recorded at Site 645. This suggests a predominance of marine organic matter at both sites. However, because of very low organic carbon and nitrogen values recorded in most samples, inorganic nitrogen contents may have influenced the C/N ratios significantly (cf., Chapter 4).

Rock-Eval pyrolysis

In Figure 41, the results of Rock-Eval pyrolysis of Site 645 sediments are plotted in a van Krevelen-type diagram (cf., Tissot and Welte, 1984). The data indicate that most of the organic matter in Plio-Pleistocene sediments (i.e., the upper 460 m) is of hydrogen-poor type III (HI values of less than 100 mgHC/gC), that means of terrigenous origin. In the Miocene interval, some higher HI values of up to 220 mgHC/gC were recorded (Figs. 41 and 42). These values may increase to 300 when measured on kerogen concentrates (cf., Fig. 24; Stein et al., 1989a). Consequently, marine organic carbon may be a more significant proportion of the total organic carbon during this time interval (although the terrigenous organic matter still remains dominant). Furthermore, increased HI values generally coincide with increased organic carbon contents (Fig. 43).

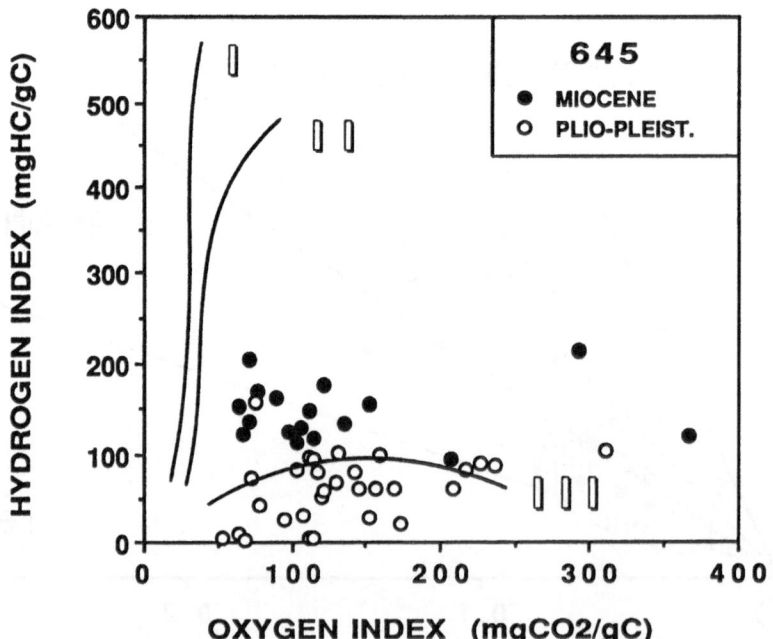

Fig. 41: Results from Rock-Eval pyrolysis: Hydrogen and oxygen index values at Site 645.

59

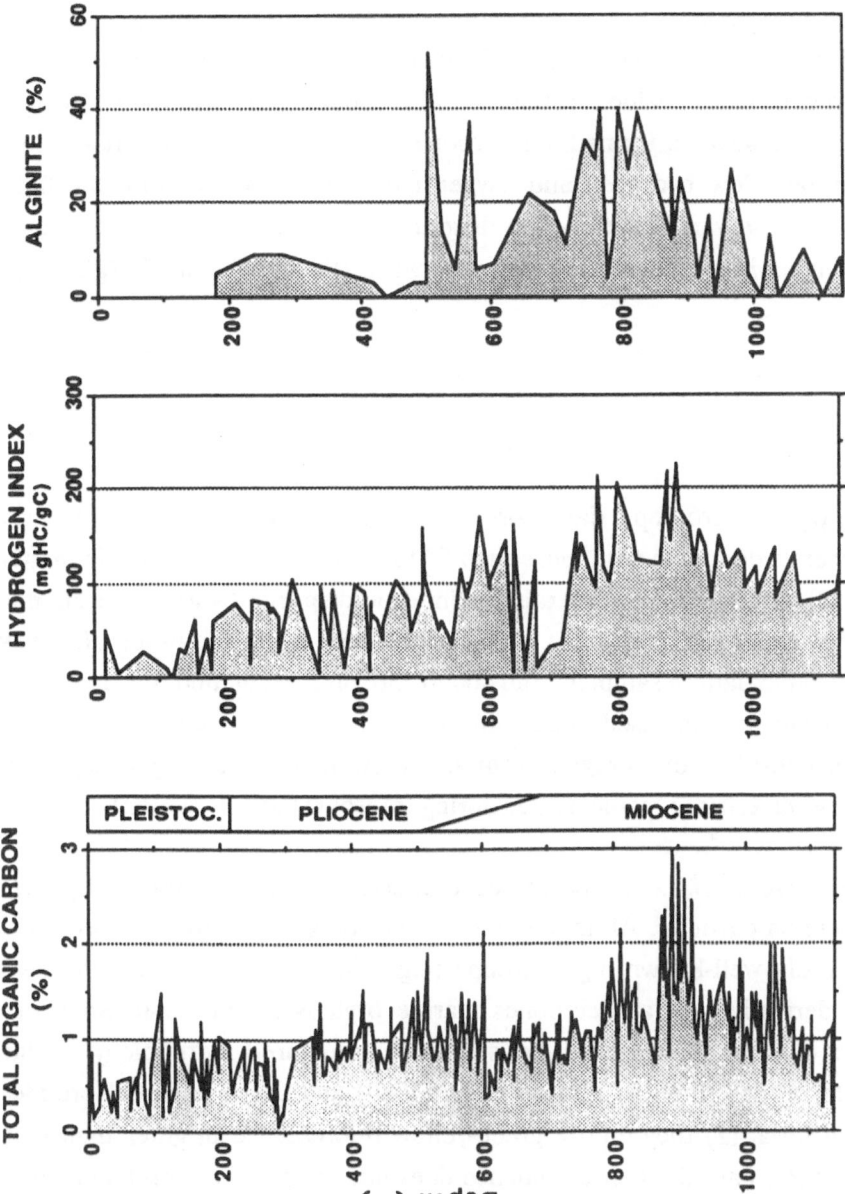

Fig. 42: Total organic carbon contents (%), hydrogen indices (mgHC/gC), and content of alginites (% of total macerals) from sediments at Site 645 (Stein et al., 1989a; Stein, 1991).

Because of the low organic-carbon contents of sediments from Sites 646 and 647, only a few Rock-Eval pyrolysis analyses were performed on bulk sediment samples, yielding very low hydrogen index values (Stein et al., 1989a). At Site 646, Rock-Eval measurements were also performed on carbonate-free samples. The Mio-/Pliocene samples display both low hydrogen and oxygen index values, falling into the field of "Kerogen type III" in a van Krevelen-type diagram (Fig. 44). The Pleistocene samples have distinctly higher hydrogen values ranging between 100 and 300 mgHC/gC (Figs. 44 and 45).

Kerogen microscopy

In general, kerogen microscopy data corroborate the Rock-Eval and C/N results; terrigenous macerals dominate with more than 50 to almost 100% the organic matter at Site 645 (Fig. 23). Among these macerals, vitrinites and sporinites are most abundant, whereas brightly reflecting inertinites and cutinites are rare (Stein et al., 1989a). Inertinite is most abundant in samples from 400 to 550 mbsf. High amounts of inertinites may either indicate an increased supply of reworked organic matter derived from erosion and redeposition of kerogen from older strata or a stronger degradation (oxidation) of the terrestrial organic matter during transport (see discussion).

Between 500 and 970 mbsf, i.e. in the Miocene interval, netlike liptinites ("Alginite" in Fig. 42) are common particles. Although these cannot be grouped together with any of the petrographically well-known algae groups (e.g., Tasmanales), these liptinites are interpreted as derivatives of autochthones marine organisms (Littke in Stein et al., 1989a), because (1) they show a characteristic, finely structured morphology similar to lamellar alginite, as described by Hutton et al. (1980) and are a potential product of benthic cyanobacteria, (2) they emit a green-yellow fluorescence brighter than that of terrigenous liptinites, and (3) their abundance does not seem to be correlated with the abundance of terrigenous organic particles.

On the organic-carbon-poor sediments of Sites 646 and 647, no kerogen microscopy study was performed.

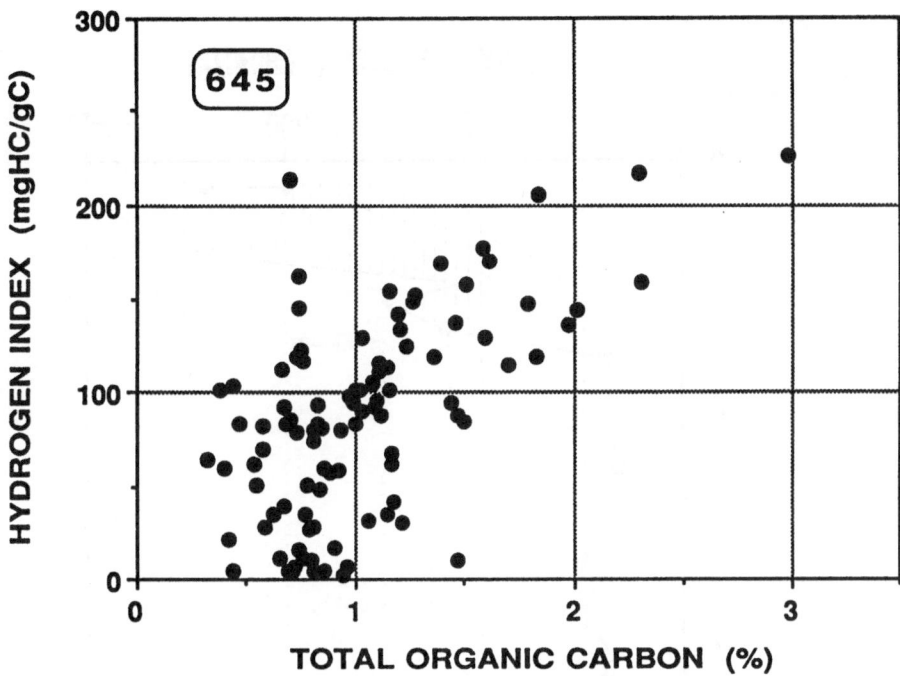

Fig. 43: Total organic carbon contents vs. hydrogen indices at Site 645.

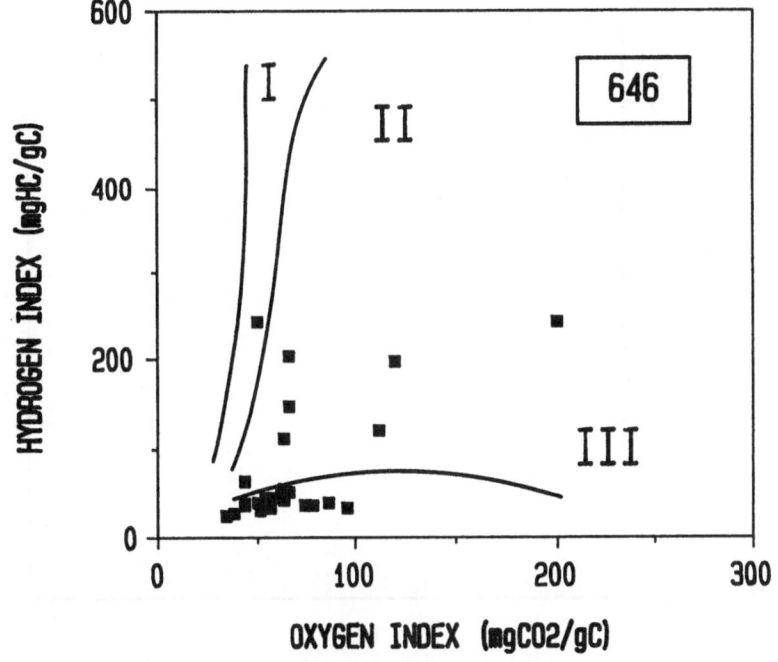

Fig. 44: Results from Rock-Eval pyrolysis: Hydrogen and oxygen index values of carbonate-free sediment samples from Site 646.

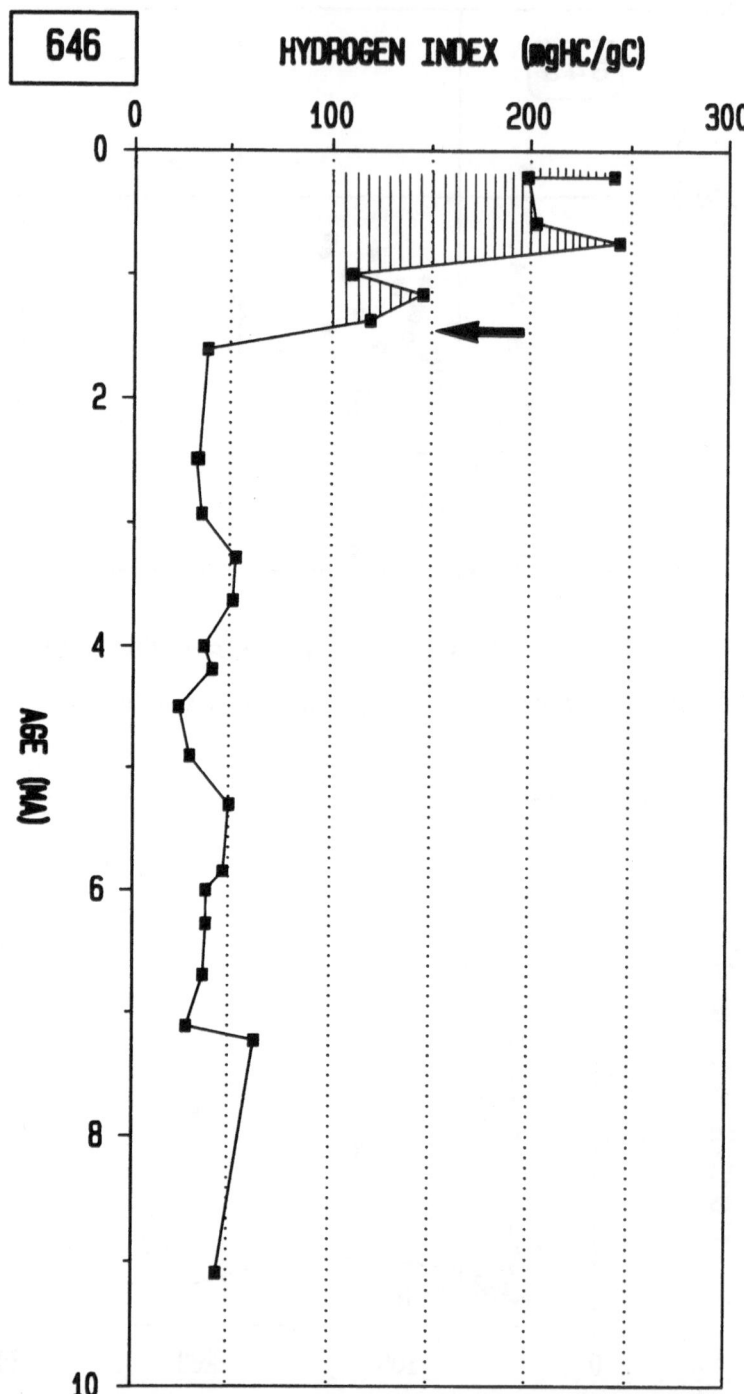

Fig. 45: Hydrogen indices (measured on carbonate-free samples) vs. depth at Site 646.

Gas chromatography and gas chromatography/mass spectrometry

Based on the results described above, ten Haven and Rullkötter (1989) have selected several samples from Site 645 for a detailed investigation of the aliphatic hydrocarbon fractions by gas chromatography and gas chromatography/mass spectrometry. According to their results, the composition of n-alkanes is dominated by long-chain homologues in all samples, especially n-$C_{25}H_{52}$, n-$C_{27}H_{56}$, n-$C_{29}H_{60}$ (most abundant), and n-$C_{31}H_{64}$ (Fig. 46). This kind of distribution pattern indicates an origin from higher plants. Other important terrigenous biological markers determined in sediments from Site 645 are oleanene and ursene (ten Haven and Rullkötter, 1989).

Thus, the study of biomarkers generally supports the results derived from kerogen microscopy, Rock-Eval pyrolysis, and elemental analyses, i.e., the dominance of terrigenous organic matter at Site 645. However, the composition of biomarkers does not support the occurrence of significant amounts of marine organic matter in the Miocene section of Site 645 as suggested from kerogen microscopy data. A possible explanation for this discrepancy given by ten Haven and Rullkötter (1989) is that at the stage of diagenesis encountered in the Miocene sediments of Site 645 (see below), marine biological markers still occur as functionalized components (e.g., sterols), whereas part of the terrigenous precursors are defunctionalized at an earlier stage and thus can already be identified in the aliphatic hydrocarbon fraction. Future investigations of the alcohol, ketone, and fatty acid fractions will evaluate this hypothesis.

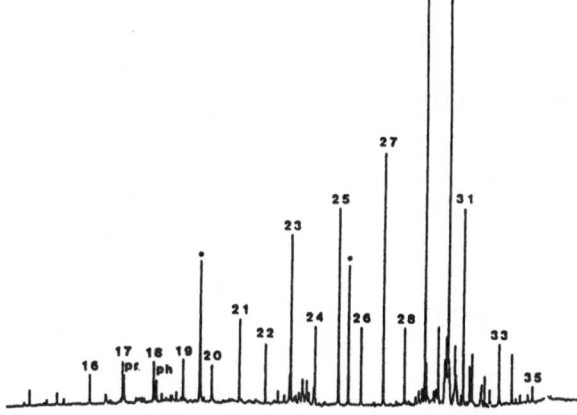

Fig. 46: Gaschromatogram of the aliphatic hydrocarbon fraction of Sample 645E-39-4/134cm (778.1 mbsf; middle Miocene). Straight chain alkanes are labeled with numbers corresponding to their carbon-atom number. Contaminants are indicated with an asterisk. Pr and Ph indicate the isoprenoid hydrocarbons pristane and phytane, respectively. (from Ten Haven and Rullkötter, 1989).

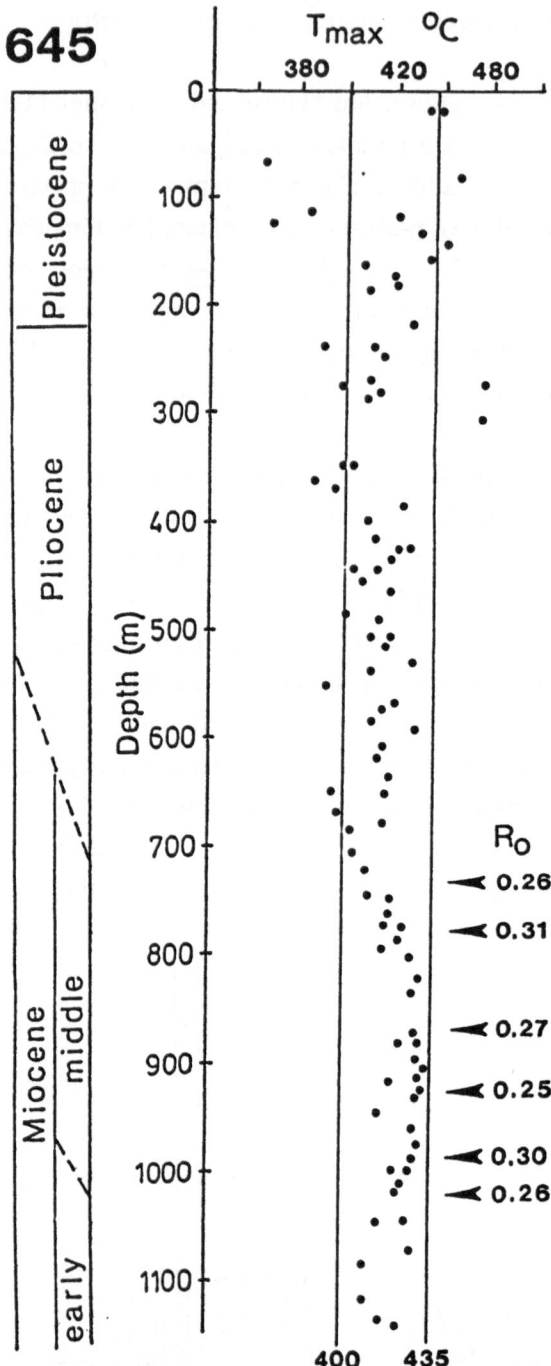

Fig. 47: Temperatures of maximum pyrolysis yield (T$_{max}$) from Rock-Eval measurements and vitrinite reflectance values (R$_0$ in %) of sediments from Site 645.

5.4.3. Maturity of organic matter

According to the reflectance values of primary vitrinite particles ranging between 0.25 and 0.31% and temperatures of maximum pyrolysis yield of less than 435°C (Fig. 47), the organic matter accumulated at Site 645 is still immature. This is also supported by the wavelength of maximum fluorescence intensity of sporinites (Littke in Stein et al., 1989a) and the strong odd-over-even carbon number predominance of the n-alkanes (ten Haven and Rullkötter, 1989). In some Plio-Pleistocene sediments from Site 645, the occurrence of inertinites and T_{max} values higher than 435°C may suggest the occurrence of reworked material from older strata (see discussion).

From sediment samples of Sites 646 and 647, no detailed information about the maturity of the organic matter is available. At Site 646, low T_{max} values of less than 428 °C indicate the low maturity of the organic matter. At Site 647, T_{max} values determined for the Oligocene interval indicate that at least in the upper 225 m of the sediment section the organic matter is immature (Stein et al., 1989a). Very high OI values derived from Rock-Eval analysis of sediments from Sites 646 and 647 (Cederberg, pers. comm. 1987) support the immaturity, too.

5.5. Discussion

5.5.1. Terrigenous organic carbon supply at Baffin-Bay Site 645 and its paleoenvironmental significance

The sediments at Site 645 are characterized by high accumulation rates of organic carbon of up to 0.2 gC cm^{-2} ky^{-1}. These values are similar, for example, to those recorded east of New Zealand (Fig. 2, DSDP-Leg 90) where organic carbon of up to 0.15 gC cm^{-2} ky^{-1} was deposited in an environment dominated by high terrigenous supply and increased surface-water productivity during glacial intervals (Dersch, 1990; Dersch and Stein, 1991). These values, however, are also similar to those recorded in the upwelling area off NW-Africa (0.1 to 0.4 gC cm^{-2} ky^{-1}; cf., Chapter 6.5.), although the depositional environment is very different. Thus, a comprehensive interpretation of the organic carbon data requires additional information concerning the composition of the organic matter (i.e., its marine and terrigenous proportions). Then, distinct differences are obvious between the data sets reflecting the different environments (see Chapter 6 for data from the upwelling area off NW-Africa).

The depositional regime at Site 645 is clearly dominated by the supply of terrigenous (organic, detricarbonate, and/or siliciclastic) material throughout the entire sediment sequence, i.e., from early Miocene to Holocene times. The dominance of terrigenous organic matter in the sediments of Site 645 is reflected in the HI values, C/N ratios, maceral composition, and GC and GC/MS data (see above). This importance of the supply of terrigenous material is not surprising because Baffin Bay is a narrow intracontinental basin close to the surrounding continents (Fig. 27). Thus, high amounts of (terrigenous) organic carbon of 0.1 to 0.2 gC cm^{-2} ky^{-1} were deposited at Site 645 (Fig. 48). These amounts are distinctly higher than those recorded in near-by open-marine environments, such as at Sites 646 and 647 (Fig. 39). Many of the terrigenous organic particles have diameters larger than 30 μm, especially in the younger sediments, indicating short-transport distances (Littke in Stein et al., 1989a). At open-marine localities that are farther away from continents or islands, terrigenous organic particles are usually distinctly smaller than 30 μm, when transportation and deposition by turbidity currents can be excluded (Stein et al., 1988).

Based on the record of accumulation rates of terrigenous organic carbon at Site 645, the evolution of paleoclimate in the Baffin Bay area through Miocene to Holocene times can be reconstructed (Fig. 48; Stein, 1991). Distinct long-term as well as short-term changes can be identified for the last 20 Ma. According to the long-term record, four major climatic intervals affected the Baffin Bay area:

Interval I (early to middle Miocene, 20 to about 14.5 Ma) is characterized by relatively high accumulation rates of terrigenous organic matter around 0.1 gC cm^{-2} ky^{-1} (Fig. 48). This is explained by a dense vegetation cover and fluvial sediment supply from Baffin Island and/or Greenland because of a more temperate paleoclimate. An increased fluvial supply is also evident from the large quantities of terrigenous clay (mainly smectites; Thiebault et al., 1989), at least for the upper part of this interval (Fig. 48; Srivastava, Arthur et al., 1987). The pollen and spores in Site 645 sediments also indicate a climate in the surrounding continents that varied within a temperate regime during early to middle Miocene (Head et al., 1989a).

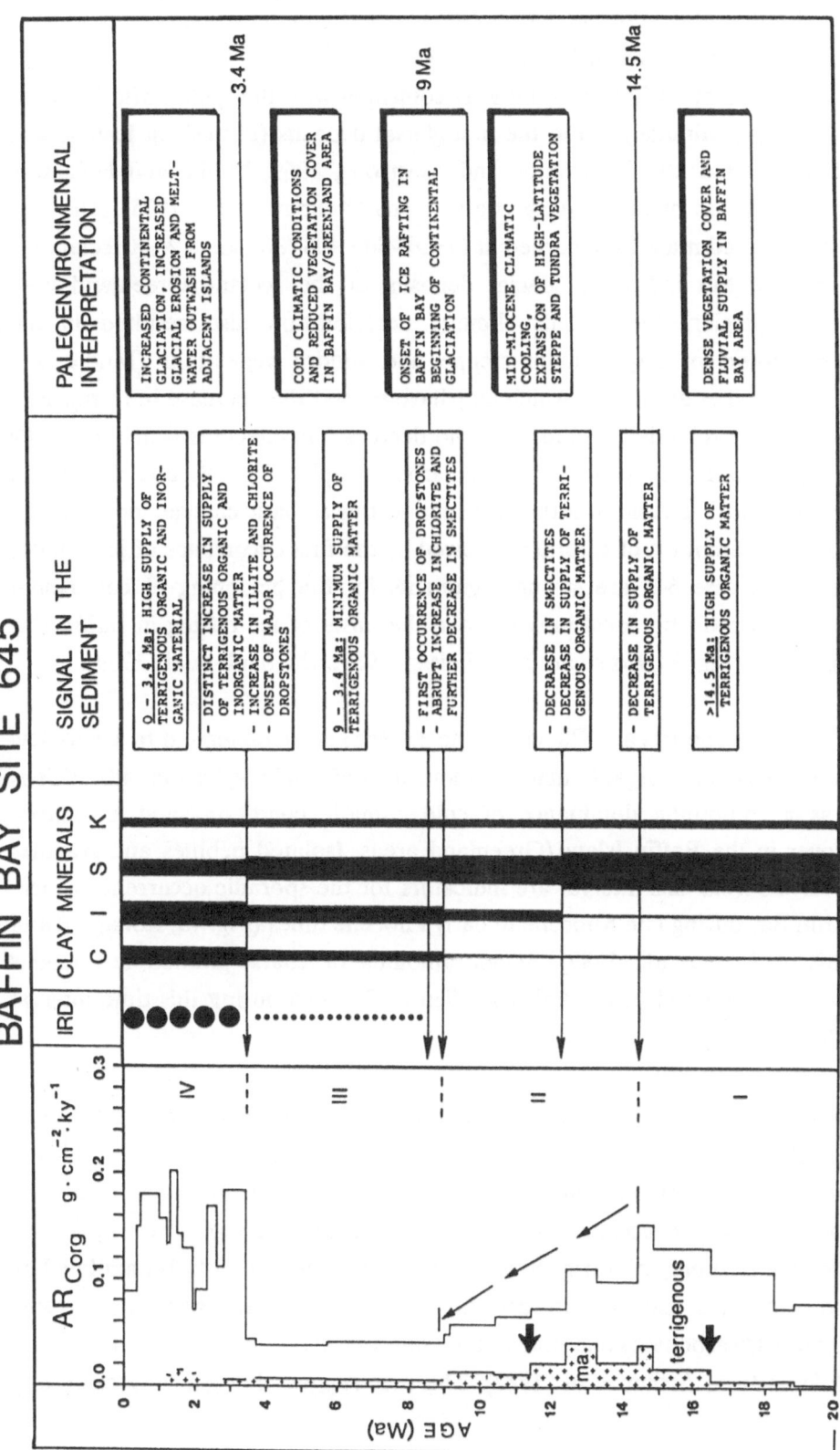

Fig. 48: Summary plot of accumulation rates of marine and terrigenous organic carbon, occurrence of ice-rafted debris (IRD) (according to Srivastava, Arthur et al., 1987), and clay mineral composition (according to Thiebault et al., 1989) at Site 645 and the paleoenvironmental interpretation. C = chlorite; I = illite; S = smectite; K = kaolinite; ma = marine.

During Interval II (middle Miocene, 14.5 to 9 Ma), accumulation rates of terrigenous organic carbon fell significantly to less than 0.04 gC cm^{-2} ky^{-1} (Fig. 48). This gradual decrease (? with steps at 14.5, 12.5 and 9 Ma) coincided with the global Mid-Miocene climatic cooling, that ultimately led to the first glacial deposits (tillites) on Iceland and in South Alaska around 10 Ma (Denton and Amstrong, 1969; Mudie and Helgason, 1983) and the formation of major Antarctic ice caps (Kennett et al., 1975). The first occurrence of ice-rafted material recorded at DSDP-Site 408 (300 km SW of Iceland) is also dated to 10.2 Ma indicating glacial deposits in the Iceland/Greenland area (Schaeffer and Spiegler, 1986). This climatic deterioration also resulted in the widespread expansion of high-latitude steppe and tundra vegetation (Mudie and Helgason, 1983; Head et al., 1989a) which may have caused the decrease in terrigenous organic matter supply recorded at Site 645. The decrease in smectites which generally parallels the decrease in terrigenous organic carbon supply (Fig. 48) and the abrupt increase of chlorite near the end of Interval II support this climatic change, too. The first occurrence of major amounts of chlorites followed by the first occurrence of dropstones in the sediments of Site 645 are strong arguments for the beginning of continental glaciation (although probably local in scale) and the onset of ice rafting in Baffin Bay between 9 and 8 Ma (Fig. 48; Korstgärd and Nielsen, 1989; Thiebault et al., 1989).

Interval III (late Miocene to early Pliocene, 9 to 3.4 Ma) is characterized by a very low contribution of terrigenous organic matter (about 0.04 gC cm^{-2} ky^{-1}; Fig. 48). This is interpreted as a continuing dominance of cold climatic conditions and a reduced vegetation cover in the Baffin Island/Greenland areas. Isolated pebbles and granules found in sediments from this interval are indicators for the sporadic occurrence of ice-rafting in Baffin Bay during late Miocene to early Pliocene times (Fig. 48; Korstgärd and Nielsen, 1989; Arthur et al., 1989). Minor amounts of coarse-grained terrigenous particles were also recorded in the Labrador Sea at Site 646 during this time interval (Wolf and Thiede, 1990).

Near 3.4 Ma, the supply of terrigenous organic carbon distinctly increased by a factor of about 5 (Fig. 48). Furthermore this climatic Interval IV (late Pliocene to Recent, 3.4 to 0 Ma) is characterized by high-amplitude short-term variations in climatically modulated terrigenous organic carbon accumulation between 0.07 and 0.20 gC cm^{-2} ky^{-1}. The important increase in supply of terrigenous organic carbon near 3.4 Ma is paralleled by an even more obvious increase in accumulation rates of terrigenous inorganic (siliciclastic and detricarbonate) material and the onset of major ice-rafted debris at Site 645 (Fig. 48; Srivastava, Arthur et al., 1987). This may suggest that glacial erosion and fluvial (meltwater) outwash of the adjacent islands caused the synchronous changes in organic and mineral matter supply. This eroded material is partly derived from pre-Quaternary sedimentary formations cropping out around Baffin Bay, as indicated by the abundance of reworked (Cretaceous and Tertiary) palynomorphs in Site 645 sediments (De Vernal and Mudie, 1989a; Hillaire-Marcel et al., 1989). Several high T_{max} values of 435 to 480 $^{\circ}$C as determined by Rock-Eval pyrolysis, and the abundance of inertinites also support the occasional occurrence of reworked, more mature organic matter in late Pliocene/Pleistocene sediments at Site 645.

The interval of high influx of terrigenous organic matter is interrupted by a short-lived period of distinctly lower accumulation rates near 2 Ma (Fig. 48). For this time interval, a distinct maximum in $\delta^{18}O$ is recorded in deep-sea cores from different parts of the world ocean, indicating a distinct glacial maximum (e.g., Shackleton and Opdyke, 1977; Keigwin, 1979; Stein, 1984; Stein and Sarnthein, 1984). This may have resulted in more extensive sea-ice cover in Baffin Bay and/or more extensive glacial ice cover on the land around Baffin Bay, causing a less dense vegetation cover and reduced supply of terrigenous (organic) matter during that time (cf., Mudie and Short, 1985).

Changes in terrigenous sediment supply between glacial/ interglacial times can also explain the negative correlation between organic carbon and (detrital) carbonate content observed in the upper 4m of the sediment sequence of Site 645 (Fig. 34). During deposition of the uppermost carbonate-rich one meter of sediment (i.e., the Holocene), high supply rates of detrital carbonates resulted in low organic carbon contents because of dilution. Based on piston core data, Aksu (1981) also decribed a high percentage of detrital carbonate of up to 50% in the ice-rafted facies within Baffin Bay. On the other hand, extensive sea-ice cover existing during the last glacial in Baffin Bay (cf., Fig. 30) prevented major ice-rafting and major supply of detrital carbonate, resulting in a relative enrichment of organic carbon. Because of a limited stratigraphic framework for Site 645, it is not possible to calculate separately accumulation rates of organic and inorganic material for glacial and interglacial intervals which would be necessary for a more detailed interpretation of glacial/interglacial fluctuations in sediment supply (cf., Chapter 5.3.3 for Site 646).

5.5.2. Marine organic carbon deposition at Site 645 and surface-water productivity in Baffin Bay

Accumulation rates of marine organic carbon are mainly controlled by the rate of supply (i.e., the surface-water productivity) and/or the preservation rate (see Chapter 2). Anoxic deep-water conditions (such as those found in the modern Black Sea; Degens and Ross, 1974) resulting in an enhanced preservation rate of organic carbon, are unlikely in Baffin Bay because bioturbation prevailed in almost the entire sediment section of Site 645 (Srivastava, Arthur et al., 1987). Rapid burial of organic carbon in turbidites can also not have been the dominant mechanism for organic carbon enrichment because turbidites are of minor importance in intervals of high organic carbon content (Srivastava, Arthur et al., 1987; Hiscott et al., 1989). Consequently, changes in accumulation rates of marine organic carbon at Site 645 are probably caused by paleoproductivity variations.

Tab. 4: Estimated mean paleoproductivity values for different time intervals of sediments from ODP-Leg 105 (Stein et al., 1989a).

Age	Sedimentation rate (cm / ky)	Porosity (%)	Wet-bulk density (g/cm^3)	Marine organic carbon (%)	Water depth (m)	Paleoproductivity (gC · m^{-2} · y^{-1})
Site 645						
late Pliocene/ Pleistocene	13.5	46	2.02	0.21 (0.13)	2000	60–90
latest Miocene/ early Pliocene	2.8	49	1.92	0.18 (0.14)	1500	50–60
late Miocene	4.8	45	2.02	0.36 0.18	1200	60–100
middle Miocene	4.8	41	2.07	0.7 (0.40)	1000	100–150
early Miocene	4.8	33	2.17	0.17 (0.13)	800	40–50
Site 646						
Pleistocene	7.9	72	1.62	0.3	3450	95
late Pliocene	7.9	67	1.67	0.3		100
early Pliocene	9.6	56	1.89	0.39		150
late Miocene (7–5.5 Ma)	9.6	46	2.11	0.37		170
late Miocene (9–7 Ma)	4.9	44	2.14	0.3		140
Site 647						
Pleistocene	4.6	63	1.72	0.22	3870	90
late Pliocene	4.6	52	1.94	0.39		160
Miocene	0.2	81	1.43	0.16		40
Oligocene	1.6	78	1.46	0.4		90
late Eocene/ earliest Oligocene	3.6	54	1.9	0.4		140
middle Eocene	0.4	43	2.13	0.22	2000–3000	90
early Eocene	2.5	39	2.19	0.11		70

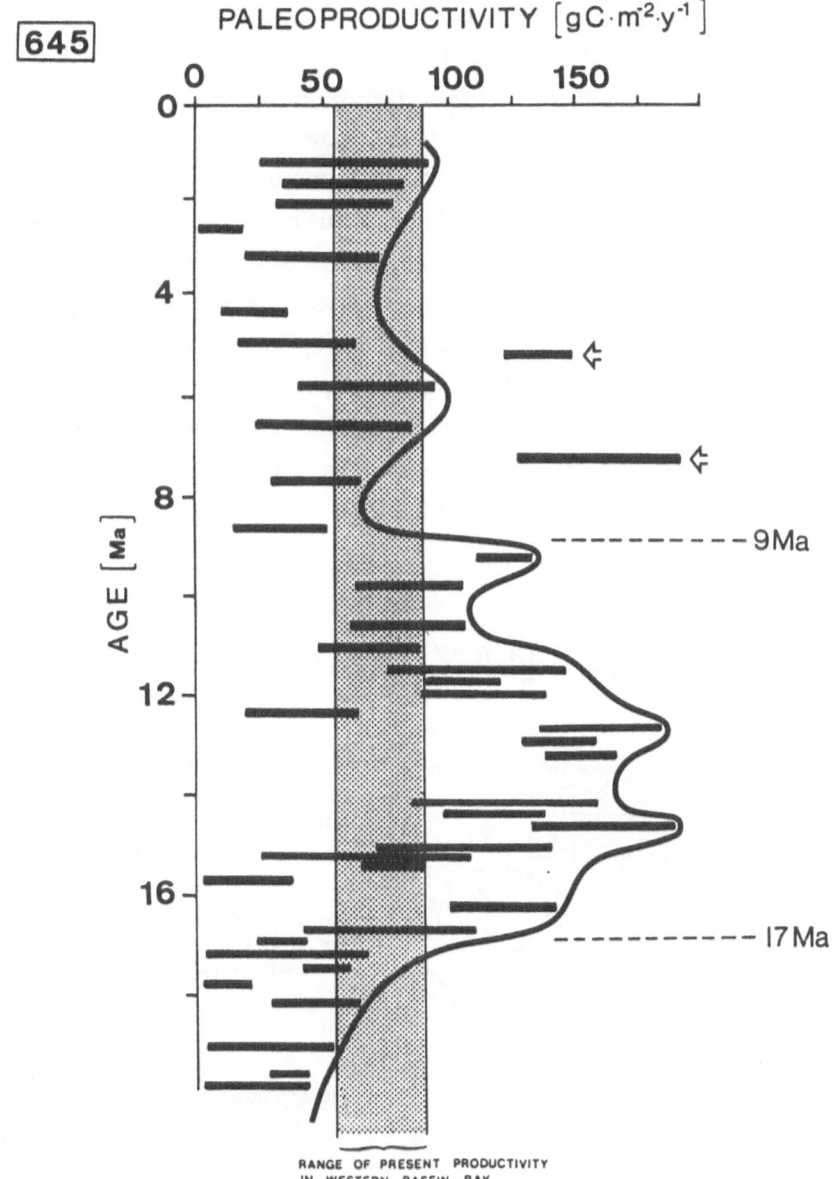

Fig. 49: Estimated paleoproductivity from marine organic carbo contents at Site 645, using equation (4). Minimum and maximum productivity values (i.e., the width of the black bar) are based on minimum and maximum estimates of marine organic-carbon contents from kerogen microscopy (Stein et al., 1989a). Range of present productivity in western Baffin Bay from Romankevich (1984). Single maximum values at 5 and 7 Ma may suggest "high-productivity events".

Generally low marine organic carbon accumulation rates of less than 0.02 gC cm^{-2} ky^{-1} suggest that Baffin Bay was an area of relatively low productivity for much of the last 20 Ma (Fig. 48). Only in the middle Miocene (between about 16.5 and 11.5 Ma), some higher values of up to 0.04 gC cm^{-2} ky^{-1} occur. Low surface-water productivity is also supported by the absence of radiolarians and low amounts of dinoflagellate cysts found in the Site 645 sediments (Head et al., 1989a; Lazarus and Pallant, 1989; Hillaire-Marcel et al., 1989). A few samples from the uppermost 22 m of the sedimentary column at Site 645, however, contained abundant dinocysts suggesting short but distinct episodes of high phytoplankton productivity in Baffin Bay during late Quaternary times (Hillaire-Marcel et al., 1989). According to Williams (1986), it is likely that the western part of Baffin Bay has a relatively high surface-water productivity at least seasonally, as today the Baffin Island Current sweeps along the ice edge and causes upwelling. Such "high-productivity events" are not documented in the long-term record of Site 645 (Fig. 48).

BAFFIN BAY (SITE 645)

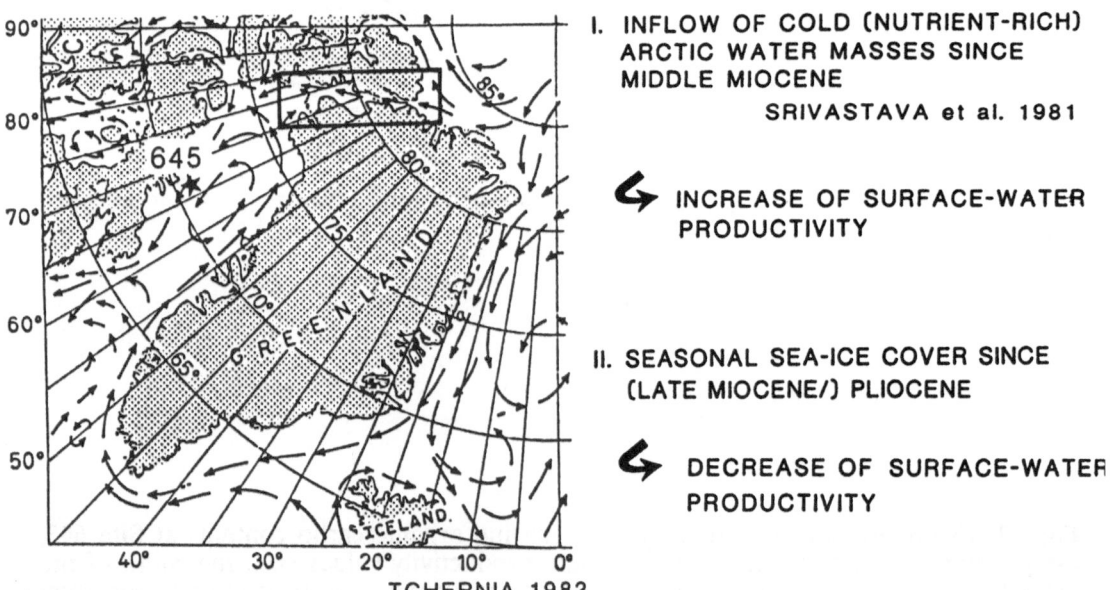

I. INFLOW OF COLD (NUTRIENT-RICH) ARCTIC WATER MASSES SINCE MIDDLE MIOCENE
SRIVASTAVA et al. 1981

↳ INCREASE OF SURFACE-WATER PRODUCTIVITY

II. SEASONAL SEA-ICE COVER SINCE (LATE MIOCENE/) PLIOCENE

↳ DECREASE OF SURFACE-WATER PRODUCTIVITY

TCHERNIA 1982

Fig. 50: Interpretation of the two major long-term changes in paleoproductivity shown in Figure 49.

Using the equation (4), paleoproductivity in Baffin Bay was estimated from the marine organic carbon content of Site 645 sediments (Fig. 49, Tab. 4; for method see Chapter 3.1; Stein, 1986a). The use of this model demonstrates that during early Miocene times productivity was very low with typical values of less than 60 gC m^{-2} y^{-1} (i.e., less than 25 gC m^{-2} y^{-1} for new production rates). Periods with productivity rates of up to 150 (200) gC m^{-2} y^{-1} (i.e., 50 (75) gC m^{-2} y^{-1} new production) occurred during the middle Miocene (about 17 to 9 Ma). Such an increase in surface-water productivity close to 16.5 Ma may have been caused by the onset of inflow of cold (nutrient-rich) Arctic Ocean-derived water masses through Nares Strait into Baffin Bay (Fig. 50; Srivastava et al., 1981). In the lower part of this interval, high fluvial supply could have acted as further nutrient source (Fig. 48, Interval I).

Near 9 Ma, productivity rates decreased; values between 30 and 90 gC m^{-2} y^{-1} (i.e., 2 to 20 gC m^{-2} y^{-1} new production) were typical for the last 9 Ma (Fig. 49, Tab. 4). These productivity values are similar to those measured in western Baffin Bay today (Fig. 49; Romankevich, 1984). Since that time, it is thought that Baffin Bay was seasonally covered by sea ice which has caused this decrease in average surface-water productivity (Figs. 28 and 30; cf., Arthur et al., 1989). The two single high-productivity peaks near 7 and 5 Ma (Fig. 49) may reflect single high-productivity events similar to those recorded by Hillaire-Marcel et al. (1989) in their high-resolution study of the uppermost 22m of the sediment sequence of Site 645.

5.5.3. Organic carbon deposition at Sites 646 and 647 and paleoceanic circulation in Labrador Sea

Because no detailed organic geochemistry study was performed on the organic-carbon-poor sediments from Sites 646 and 647, it is not possible to differentiate precisely the terrigenous and marine proportions of the organic matter. Low C/N ratios may point to the presence of marine organic carbon in the sediments of Sites 646 and 647 (Fig. 40) although these ratios have to be interpreted very cautiously (see Chapter 4). The relatively high hydrogen index values of the Quaternary sediments from Site 646 (Fig. 45) also indicate a marine origin of the organic matter at least for this time interval. The occurrence of dinocysts, acritarchs, diatoms, and radiolarians and their changes through time in the sediments of Site 647 and, especially, Site 646 (Lazarus and Pallant, 1989; Monjanel and Baldauf, 1989; de Vernal and Mudie, 1989b) suggest that changes in surface-water productivity have influenced the sediment composition in the Labrador

74

Sea. Correlations between organic carbon content and dinocyst abundance in upper Quaternary sediments (see below) also support that organic carbon variations give information about changes in paleoceanographic conditions. It has to be considered, however, that part of the organic matter certainly is of terrigenous origin because (1) these sediments contain significant amounts of pollen and spores (Aksu et al., 1989; de Vernal and Mudie, 1989b) and (2) in open-ocean oxic environments (such as the Labrador Sea) terrigenous organic matter becomes relatively enriched because of its higher resistance against oxic decomposition in comparison to marine organic matter (e.g., Tissot et al., 1979; Waples, 1983; cf., Chapter 2.3.). This may also explain the very low hydrogen index values of Miocene/ Pliocene sediments at Site 646 (Fig. 545)

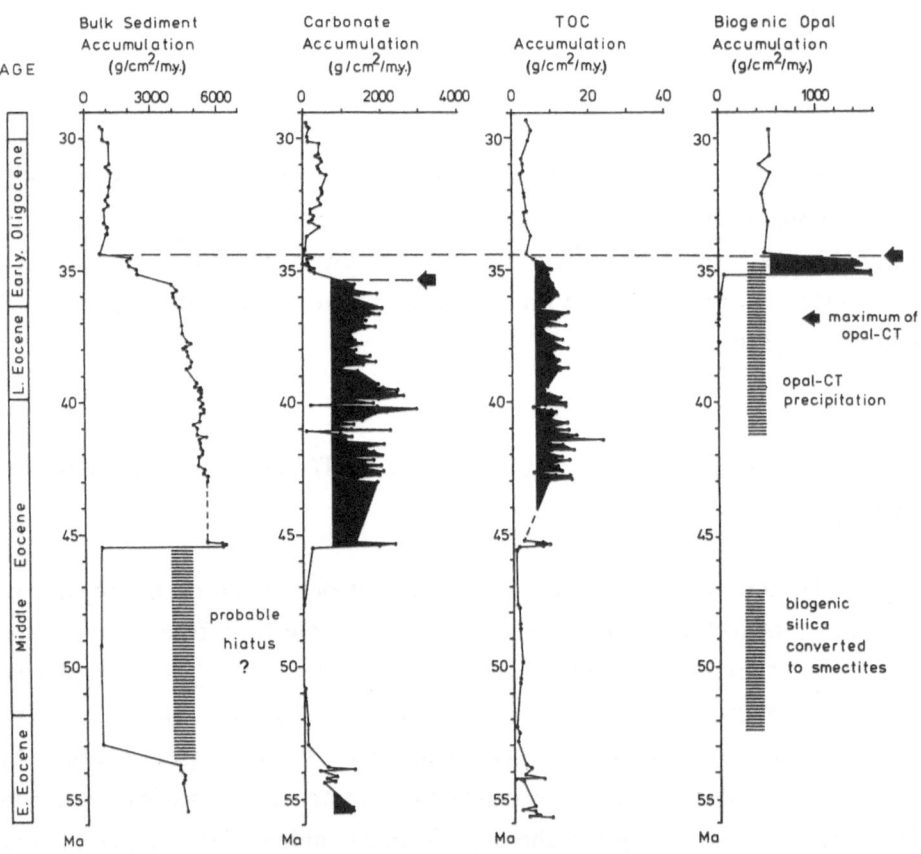

Fig. 51: Accumulation rates of bulk sediment, carbonate, total organic carbon, and biogenic opal vs. age (29.5 to 55.2 Ma) at Site 647 (from Bohrmann and Stein, 1989).

Changes in organic carbon accumulation in the Eocene Labrador Sea

A major change in organic carbon accumulation occurs in Eocene to lower Oligocene sediments at Site 647. Between 43 and 35 Ma, a maximum in accumulation rates of organic carbon and carbonate was recorded, which coincides with increased deposition of biogenic opal as indicated by maximum accumulation rates of biogenic opal and maximum formation of opal-CT (Fig. 51; Bohrmann, 1988; Bohrmann and Stein, 1989; Lazarus and Pallant, 1989). These signals are interpreted as indicators for increased productivity. Based on organic carbon data, mean surface-water productivity may have reached rates of about 150 gC m^{-2} y^{-1} (Tab. 4), i.e., rates which are distinctly higher than those measured in normal open-ocean environments, but, which are still lower than those recorded in true high-productivity areas such as coastal upwelling regions (see Chapter 6). This increase in productivity was accompanied by a cooling of surface waters (Srivastava, Arthur et al., 1987). The first inflow of cold (and nutrient-rich) surface-water masses from the Norwegian-Greenland Sea and the Arctic Ocean into the North Atlantic occurred during that time (e.g., Thiede and Eldholm, 1983). The first cool, dense Norwegian-Greenland Sea overflow through the Faeroer-Shetland Channel into the eastern North Atlantic basins is also dated near the Eocene/Oligocene boundary by several authors (e.g., Miller and Tucholke, 1983; Roberts, Schnitker et al., 1984). Although the Labrador Sea was relatively isolated from the eastern North Atlantic basins, the cold deep water may have flown through the Charlie Gibbs Fracture Zone into the Labrador Sea (Tucholke and Mountain, 1986; Arthur et al., 1989). As upwelled nutrient-rich water, it may have fertilized the surface layer, causing enhanced plankton production (cf., Bohrmann and Stein, 1989).

Besides this paleoceanographic event, it has to be considered that the Eocene was a time interval of warm climatic conditions and intensive chemical weathering (Frakes, 1979; Valeton, 1983). This probably caused higher silicate mobilization on land and a higher influx of dissolved silica to the ocean (Bohrmann and Stein, 1989). This may explain the maximum occurrence of biogenic opal recorded in the North Atlantic during early/middle Eocene times (e.g., Ehrmann and Thiede, 1985; Thein and von Rad, 1987).

Late Neogene to Quaternary changes in organic carbon accumulation

The organic carbon record of Site 646 gives important information about changes in paleoceanic conditions in the Labrador Sea during the last about 10 Ma (Fig. 52):

76

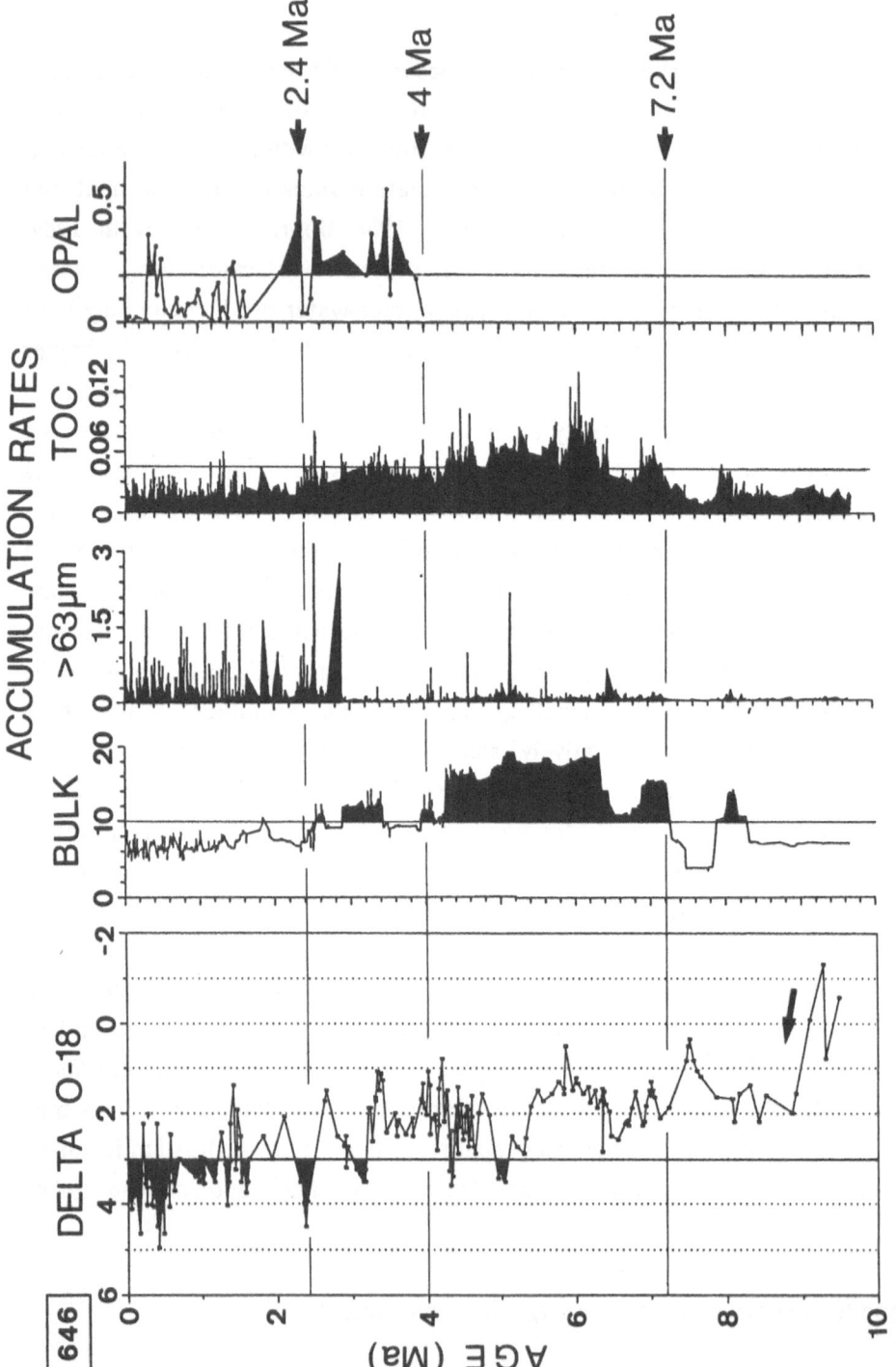

Fig. 52: $\delta^{18}O$ values ($^o/_{oo}$ PDB) of planktonic foraminifers (Aksu and Hillaire-Marcel, 1989) and accumulation rates (g cm^{-2} ky^{-1}) of bulk sediment, terrigenous coarse fraction, total organic carbon (Wolf and Thiede, 1990), and biogenic opal (Bohrmann, 1988) at Site 646.

On a long-term scale, two major changes in organic carbon accumulation rates occur: Near 7.2 Ma, these rates distinctly increased, whereas at about 2.4 Ma a decrease is obvious (Fig. 52; Wolf and Thiede, 1990). The increase in organic carbon accumulation at about 7.2 Ma coincided with a drastic increase in bulk accumulation rates and a major benthic faunal change, interpreted as the onset of significant amounts of Denmark Strait Overflow Water during that time (Arthur et al., 1989; Bohrmann et al., 1990). The increased spill-over of deep waters from the Norwegian-Greenland Sea was possible because significant cooling of surface waters occurred at high latitudes and the Greenland-Scotland Ridge had subsided sufficiently. The $\delta^{18}O$ record of planktonic foraminifera may also indicate a cooling of surface-water temperature in Labrador Sea (Fig. 52; Aksu and Hillaire-Marcel, 1989). Dinocyst assemblages, planktonic foraminifers, and calcareous nannoplankton also indicate cool to temperate surface waters for the upper Miocene at Site 646 (Aksu and Kaminski, 1989; Head et al., 1989b; Knüttel et al., 1989). These cold and nutrient-rich water masses (the Proto-East Greenland Current ?) may have resulted in increased surface-water productivity (Tab. 5) causing increased accumulation of organic carbon at Site 646 (Fig. 52).

This interpretation of a late Miocene East Greenland Current differs from that of Bohrmann et al. (1990) who interprete the first occurrence of biogenic opal at Site 646 near 4 Ma (Fig. 52) as the onset of the East Greenland Current. If a change in silica preservation/dissolution can be excluded (Bohrmann, 1988; Bohrmann et al., 1990), this biogenic opal signal may reflect an intensification or cooling of the East Greenland Current system near 4 Ma. However, it has to be considered that the low hydrogen index values of the Miocene/ Pliocene sediments (Fig. 45) can also be taken as a hint to the presence of major amounts of terrigenous organic matter. Thus, the organic carbon data do not allow a clear interpretation for this time interval.

At about 2.4 Ma, the accumulation rates of organic carbon and biogenic opal decreased at Site 646 (Fig. 52). Furthermore, this major change is almost contemporaneous with a distinct increase in $\delta^{18}O$ (Fig. 52), indicating the development of major Northern Hemisphere Glaciation with its typical glacial/interglacial cycles characterizing the climate of the last 2.4 Ma (e.g., Shackleton et al., 1984). These signals are interpreted to represent a distinct decrease in long-term surface-water productivity caused by a closed sea-ice cover dominating (seasonally) in the northern Labrador Sea during the last 2.4 Ma at least during glacial intervals (see below). Large-scale ice-rafting probably occur in the Labrador Sea prior to this event as documented by the increase of the amount of coarse-grained rock fragments in Site 646 sediments near 4 Ma (Wolf and Thiede, 1990). The increase of accumulation rates of the > 63 μm fraction at about 2.9 Ma (Fig. 52) is interpreted as a further strengthening of ice-rafting activity (Wolf and Thiede, 1990).

In order to better understand the relationships between surface-water productivity and glacial/interglacial climates, several sediment intervals at Site 646 were studied in detail. In Figures 53 and 54, organic carbon contents, abundances of dinocysts, and δ^{18}O data from Cores 646A-1H and 646A-2H are shown for the paleoenvironmental history of the last 240,000 years (i.e., δ^{18}O stages 7 to 1). According to these records, high amounts of (marine; cf., Fig. 45) organic carbon and high concentrations of dinocysts clearly correlate with interglacial stages, whereas the glacial stages 2 and 6 are characterized by low organic carbon values and low concentrations of dinocysts (Fig. 53).

In order to interpret the organic carbon data in terms of changes in rates of supply, the percentages of organic carbon were transformed into accumulation rates using the detailed sedimentation rate record from the publication of Hall et al. (1989) and mean physical property data from that of Srivastava, Arthur et al. (1987). High accumulation rates of organic carbon occur in the interglacial stages 1, 5, and 7, low rates occur in the glacial stages 2 and 6 (Fig. 54). Based on the dinocyst data (Fig. 53; Aksu et al., 1989) and hydrogen indices (Fig. 45), these accumulation rates are indicators for marine organic carbon flux. It has to be considered, however, that the pollen and spores records also indicate the presence of significant amounts of terrigenous organic carbon in these sediments (Aksu et al., 1989).

The accumulation rates of organic carbon and the paleoproductivity rates estimated from the organic carbon data (Fig. 54) indicate that during interglacial times surface-water productivity was higher than in the glacial situation (70-170 versus 30-70 gC m^{-2} y^{-1}, or 10-60 versus 2-10 gC m^{-2} y^{-1} for new production values). These differences can be explained by the different paleoceanographic/paleoclimatic situations. A closed sea-ice cover during glacial intervals on the average resulted in a reduced productivity in the surface waters (Situation A in Fig. 55); open-ocean conditions with decreased sea-ice cover led to increased surface-water productivity (Situation B in Fig. 55). Maximum productivity probably was reached when the Site 646 position was close to the sea-ice edge (i.e.., during the winter season of interglacials or during glacial/interglacial transitions). Distinct phytoplankton blooms in front of the sea-ice edge were decribed by several authors for modern polar/subpolar marine environments (e.g., Sakshaug and Holm-Hansen, 1984; Wilson et al., 1986; Nelson et al., 1989).

79

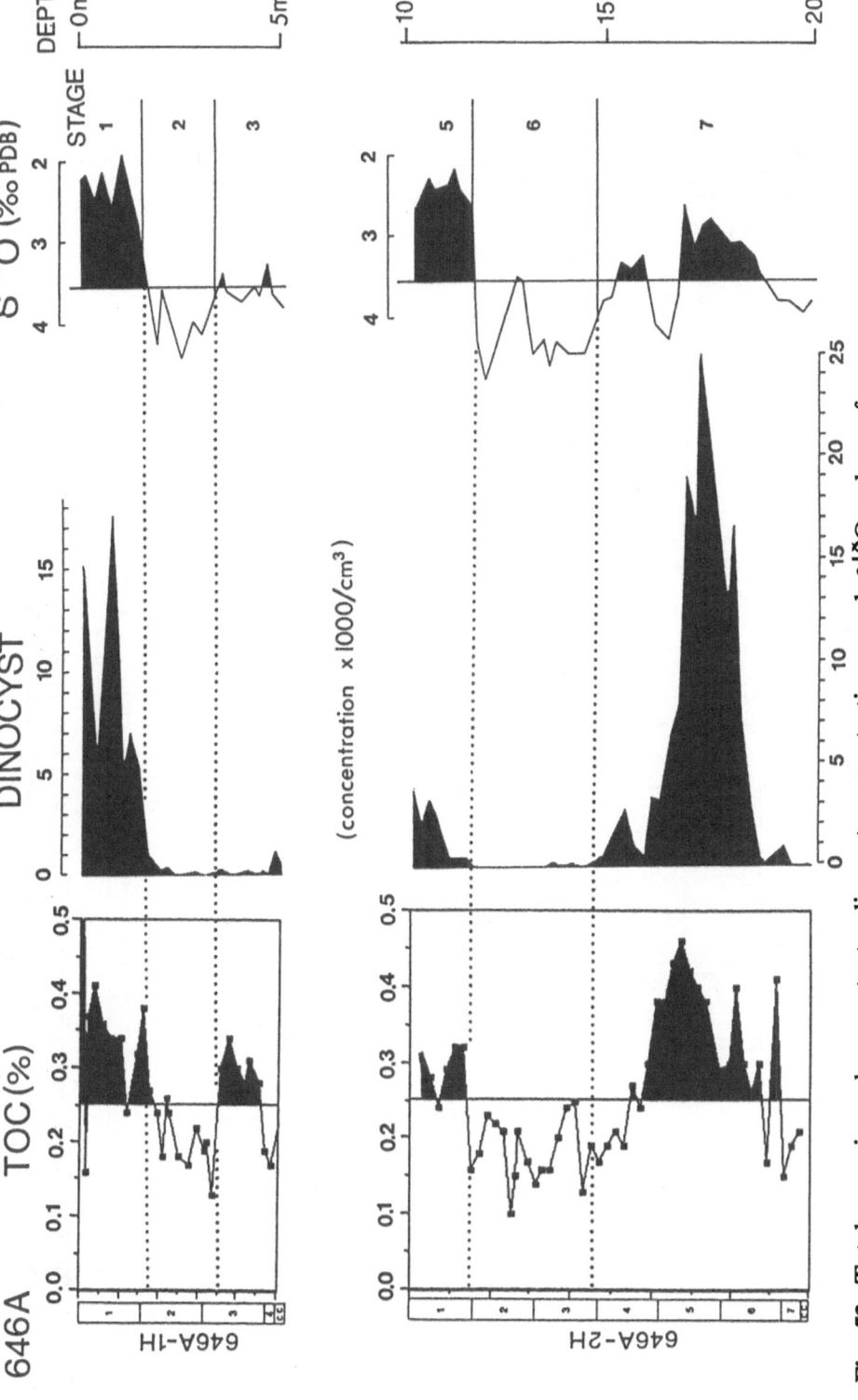

Fig. 53: Total organic carbon contents, dinocyst concentrations, and $\delta^{18}O$ values of sediments from Cores 646A-1H and 646A-2H (i.e., $\delta^{18}O$ stages 1 to 7). Organic carbon data Core 646A-1H from Macko (1989), dinocyst and stable isotope data from Aksu et al., 1989.

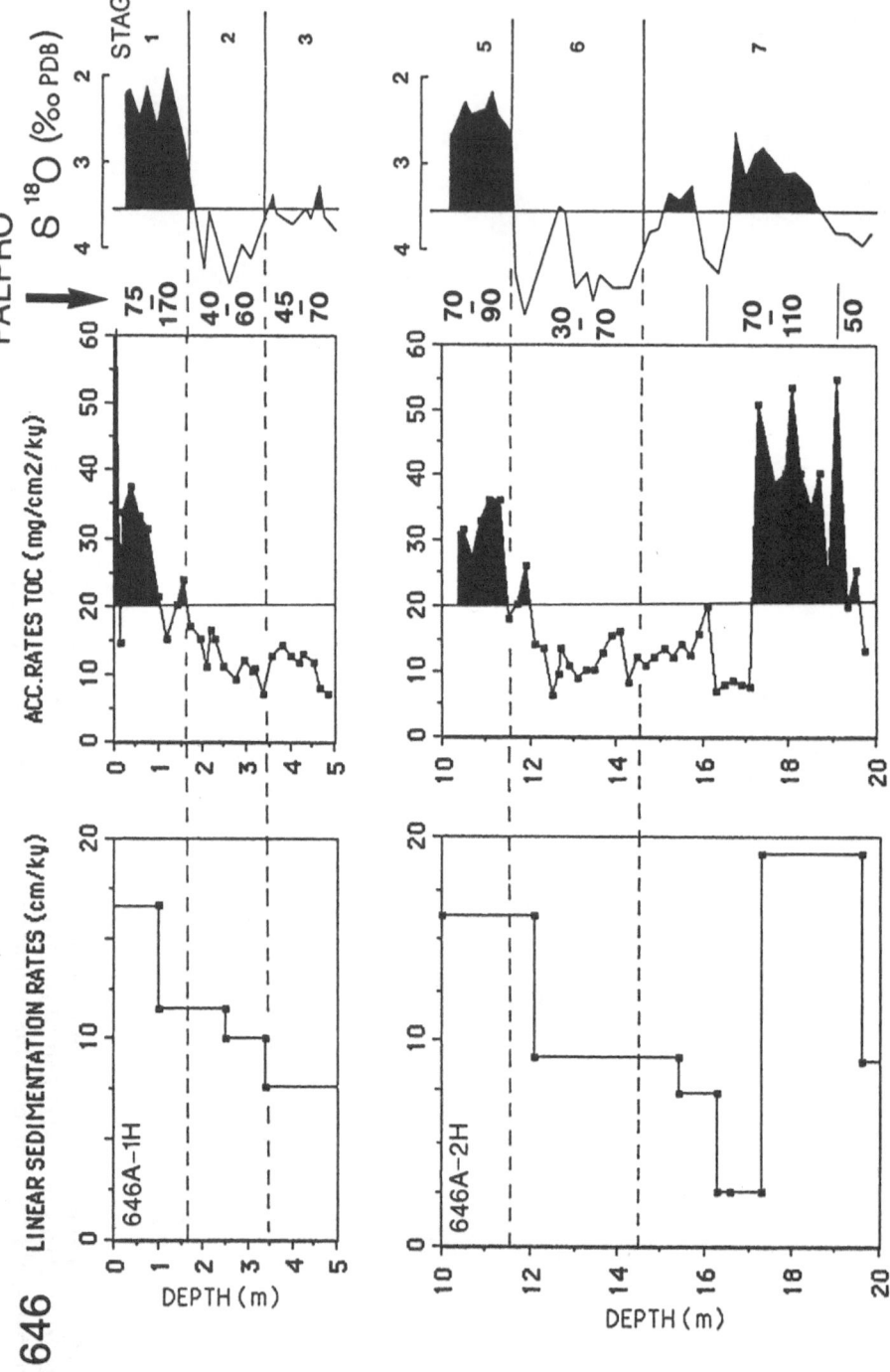

Fig. 54: Sedimentation rates (from Hall et al., 1989), accumulation rates of total organic carbon, and $\delta^{18}O$ values (cf., Fig. 53). Estimates of paleoproductivity values (in gC m^{-2} y^{-1}), calculated using equation (4).

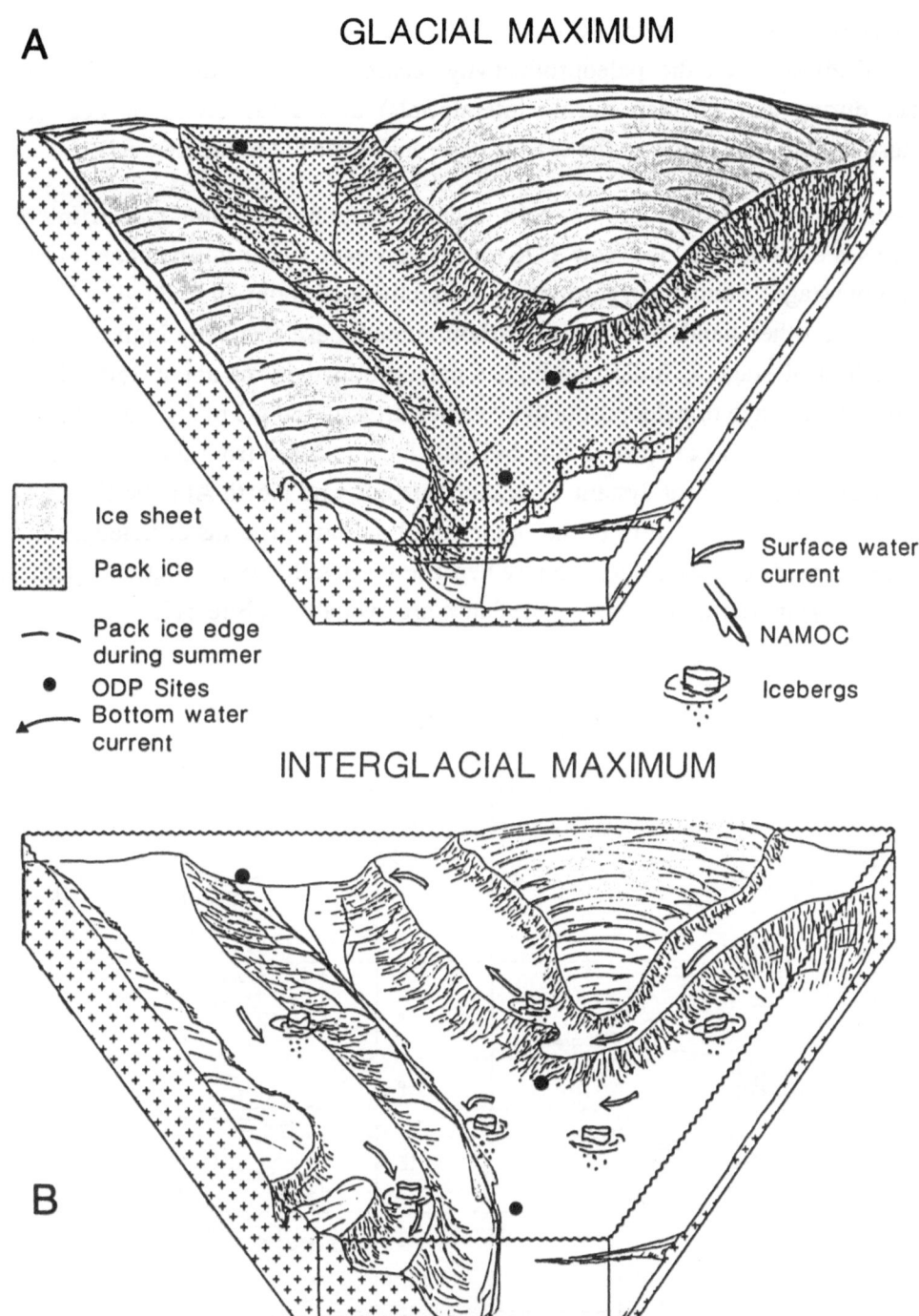

Fig. 55: Depositional model for southern Baffin Bay and Labrador Sea area with location of Sites 645, 646, and 647 (based on CLIMAP (1981) data, from Stax, 1989; Stein and Stax, 1991). A. Glacial situation. B. Interglacial situation.

82

If changes in sea-ice cover distinctly influenced the surface-water productivity as decribed above, then the paleoproductivity record should be different (i.e., possibly higher during glacials than during interglacials) at central Labrador Sea Site 647 because this site was closer to the sea-ice edge during glacial times (Fig. 55).

Indeed, no clear correlation between organic carbon data and glacial/interglacial intervals is obvious (Fig. 56). Low TOC values, however, occur during peak interglacials of isotopic stages 5 and 7, suggesting low productivity of about 40 to 80 gC m^{-2} y^{-1} (Fig. 56). On the other hand, the organic carbon contents are distinctly higher during cold interstadials of stage 7, indicating higher paleoproductivity of about 150 gC m^{-2} y^{-1}. During glacial stage 6, paleoproductivity values around 100 gC m^{-2} y^{-1} are typical (Fig. 56). The more complex record at Site 647 may have resulted from the varying influence of the Greenland Current System, Labrador Current, and North Atlantic Drift (Aksu et al., 1989). This interpretation of the organic carbon data in terms of paleoproductivity, however, should be used as preliminary because no detailed data of quality parameters of the organic matter and dinocyst abundance are available at Site 647.

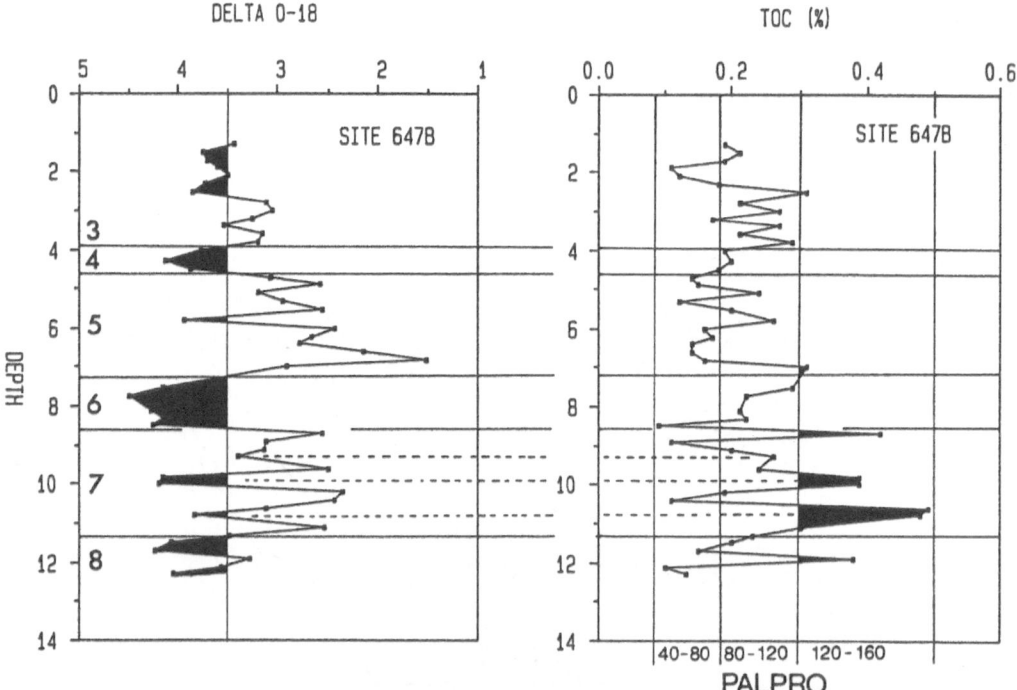

Fig. 56: δ^{18}O values (*N. pachyderma sin.*) (Aksu et al., 1989) and total organic carbon contents (Macko, 1989) of Pleistocene sediment samples from Site 647.
PALPRO = paleoproductivity (gC m-2 y-1).

5.6. Conclusions

The results of the investigation of organic carbon deposition in Baffin Bay and Labrador Sea provide important information about the paleoenvironmental evolution in high northern latitudes during Cenozoic times. Furthermore, the Baffin Bay data can be used as example/model for organic carbon enrichment in a narrow intracontinental basin characterized by high terrigenous sediment supply and fully oxic deep-water conditions. In particular, the results can be summarized as follows:

Baffin Bay

(1) The Neogene and Quaternary sediments of Baffin Bay Site 645 are characterized by relatively high organic carbon contents of 0.5 to 3%, with the maximum values occurring during middle Miocene times. Organic carbon enrichment in Baffin Bay was mainly controlled by increased supply of terrigenous organic matter throughout the last 20 Ma. Two distinct maxima were identified: (i) A middle Miocene maximum possibly reflecting a dense vegetation cover on and fluvial sediment supply from adjacent islands, decreasing during late Miocene and early Pliocene times because of expansion of tundra vegetation due to global climatic deterioration and (ii) a late Pliocene/Pleistocene maximum probably caused by glacial erosion and meltwater outwash.

(2) Significant amounts of marine organic carbon were deposited in Baffin Bay during middle Miocene times suggesting higher surface-water productivity triggered by the inflow of cold and nutrient-rich Arctic water masses. The decrease in average surface-water productivity to values similar to those of the modern Baffin Bay occurred during the late Miocene and were probably caused by the development of a seasonal sea-ice cover.

Labrador Sea

(3) A first maximum in accumulation rates of organic carbon and biogenic opal and the presence of opal-CT in late Eocene/early Oligocene sediments of Site 647 suggest increased surface-water productivity in the Labrador Sea probably induced by the first inflow of cool, more nutrient-rich water masses from the Norwegian-Greenland Sea into the Atlantic Ocean.

(4) The major increase in organic carbon accumulation recorded at Site 646 near 7.2 Ma may indicate enhanced productivity triggered by the onset of the cold East-Greenland Current system. However, increased flux rates of terrigenous organic matter cannot be excluded.

(5) Near 2.4 Ma, parallel to the development of major Northern Hemisphere Glaciation, accumulation rates of organic carbon and biogenic opal distinctly decreased at Site 646. This is thought to be due to a reduced long-term surface-water productivity because of the development of closed seasonal sea-ice cover in northern Labrador Sea.

(6) According to the organic carbon data, average surface-water productivity was significantly higher during interglacials at northern Labrador-Sea Site 646, whereas at central Labrador-Sea Site 647 productivity was higher during glacials. These differences are explained by increased productivity close to a sea-ice margin position and decreased productivity during times of closed sea-ice cover.

6. Accumulation of organic carbon in the upwelling area off Northwest Africa (ODP-Leg 108)

6.1. Introduction

During ODP-Leg 108, drilling locations were occupied at 12 sites in the eastern Equatorial Atlantic and along the Northwest African continental margin (Fig. 57). Principal objectives of Leg 108 were (1) to study the development and fluctuation in the position and intensity of the Intertropical Convergence Zone (ITCZ) and the low-level wind flow for better understanding the history of aridity in Africa, (2) to study the history of surface- water and deep-water circulations, (3) to study the evolution of coastal and equatorial upwelling in relation to climatic change, and (4) to document changes in global ice volume and low-latitude temperature to determine the climatic linkages between low and high latitudes (Ruddiman, Sarnthein, et al., 1988).

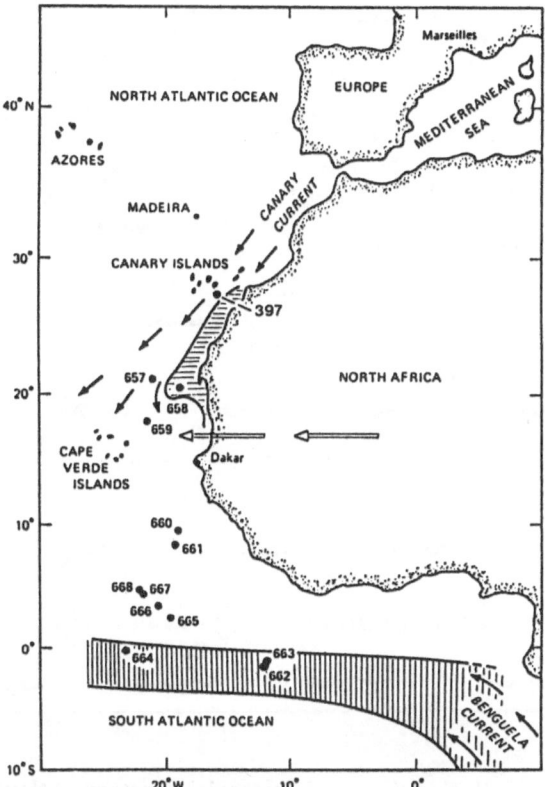

Fig. 57: Location of ODP-Sites 657-665 and DSDP-Site 397. Hatched area marks upwelling off NW-Africa. Black arrows indicate meridional trade winds, open arrows indicate mid-tropospheric Saharan-Air Layer.

Some holes were drilled in upwelling and divergence zones (Figs. 57 and 58) characterized by increased surface-water productivity, i.e., in environments with a high potential for the formation of organic-carbon-rich sediments (see Fig. 3 and below). These sites are of special interest for studies on mechanisms controlling the accumulation of organic carbon in marine environments. This study concentrates on changes in quantity and composition of organic matter deposited at Sites 657 and 659 (non-upwelling), 658 (coastal upwelling), 660 (North Equatorial Divergence Zone), and 662 (Equatorial upwelling), with a special interest to the coastal upwelling-Site 658.

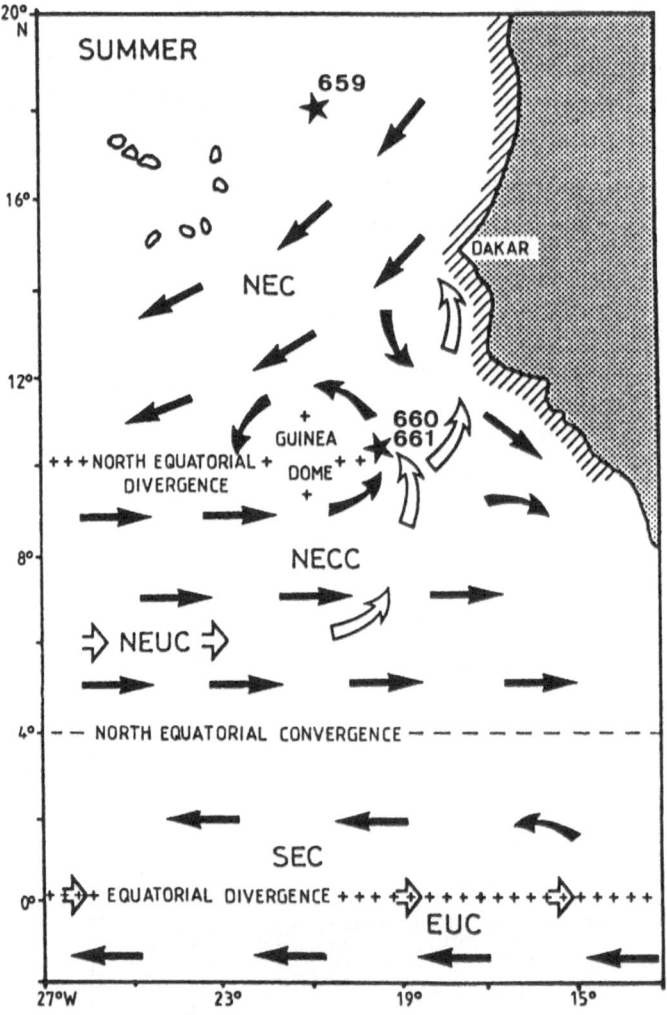

Fig. 58: Surface-water circulation in the Equatorial Eastern Atlantic. NEC = North Equatorial Current; NECC = North Equatorial Counter Current; NEUC = Northern Equatorial Undercurrent; SEC = South Equatorial Current; EUC = Equatorial Undercurrent. (from Tiedemann, 1985).

6.2. Recent and subrecent environment in the eastern tropical and subtropical Atlantic Ocean

Climate in Northwest Africa and terrigenous sediment supply

Off Northwest Africa, two major mechanisms are controlling the terrigenous sediment supply to the Atlantic Ocean. The first source, the eolian dust supply, dominates the modern depositional environment north of 15°N (i.e., off the arid to semi-arid Saharan/Sahelian area) and in offshore positions. The main transport of dust occurs in the Saharan Air Layer (SAL), a mid-tropospheric zonal wind system in 1000 to 5000 m altitude (Carlson and Prospero, 1977).

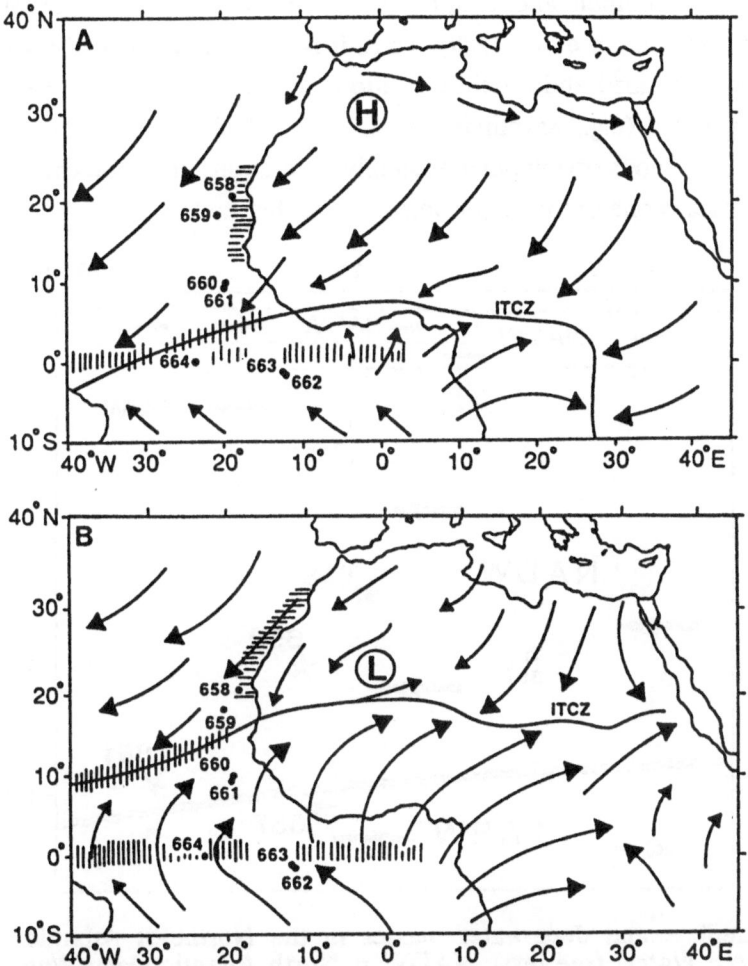

Fig. 59: Average low-level wind circulation, Intertropical Convergence Zone (ITCZ), and oceanic divergence/upwelling zones (hatched areas) in northern winter (A) and summer (B) (from Ruddiman et al., 1989, supplemented).

These dust outbreaks follow the temperature gradient linked to the ITCZ, which today reaches its northernmost position near 18°N during northern summer. Dust transport of secondary importance occurs by the NE-Trades near the surface below 1000 m altitude (Fig. 57; Tetzlaff and Wolter, 1980; Sarnthein et al., 1981). The second source, fluvial sediment supply, today is only important in the shelf/continental slope areas south of 15°N where climate is humid.

Because the sediments transported by different mechanisms and from different source areas have very different and specific, climate-dependant characteristics (such as clay mineral composition, grain size, and terrigenous organic matter), it is possible to reconstruct changes in source of terrigenous material and paleoclimate in the past from sediment data (e.g., Leinen and Heath, 1982; Sarnthein et al., 1982; Stein, 1986c; Chamley, 1989; Tiedemann et al., 1989). Thus, it is well known that distinct changes in paleoclimate between arid and humid conditions in Northwest Africa occurred during Plio-Pleistocene times (e.g., Sarnthein et al., 1981, 1982; Stein, 1985a, 1986c). The fluctuations in terrigenous organic-matter supply in relationship to the climatic evolution in Northwest Africa is one of the major objectives of this study.

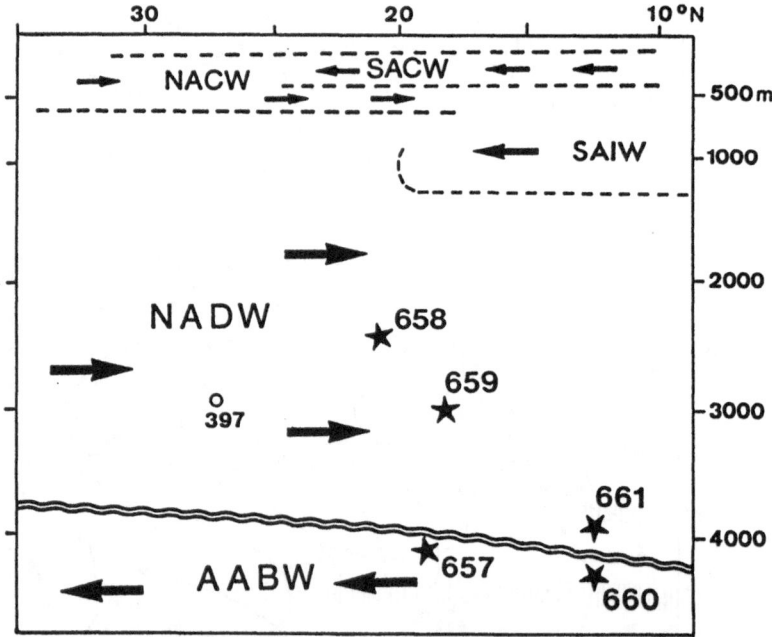

Fig. 60: Intermediate and deep-water masses in the Northeast Atlantic. AABW = "Antarctic Bottom Water" (see text); NADW = North Atlantic Deep Water; SAIW = Subantarctic Intermediate Water; NACW = North Atlantic Central Water; SACW = South Atlantic Central Water (after Sarnthein et al., 1982 and references therein) with locations of Sites 657 - 661.

Surface-water circulation and upwelling

The surface-water circulation in the eastern equatorial Atlantic Ocean is dominated by the North and South Equatorial Currents, which flow from east to west and are fed by the eastern boundary currents (i.e., the Canary and Benguela Currents) along the western coast of Africa (Figs. 57 and 58). These surface-water currents are controlled by the meridional atmospheric circulation, i.e., the NE- and SE-Trades. These Trades also trigger the position and intensity of upwelling zones (coastal upwelling, Equatorial and North-Equatorial Divergence Zones; Fig. 59). As the ITCZ and the trade wind belts shift seasonally, upwelling activities (and thus high-productivity belts) also change with season and latitude (Fig. 59): South of about $20^{\circ}N$ upwelling occurs during winter, whereas north of $25^{\circ}N$ upwelling occurs during summer times; off Cap Blanc (i.e., between about 20 and $23^{\circ}N$), upwelling persists throughout the year (cf., Speth et al., 1978; Smith, 1983). Intense coastal upwelling, however, is not always accompanied by high productivity, as indicated by the recent situation off Northwest Africa: Upwelling occurs between $12^{\circ}N$ and $33^{\circ}N$, whereas high oceanic productivity is only observed south of about $23^{\circ}N$, i.e., in areas where the nutrient-rich Subantarctic Intermediate Water (SAIW) and the South-Atlantic Central Water (SACW) are the source of the upwelled water (Fig. 60). On the other hand, the less nutrient-rich North Atlantic Central Water (NACW) is the source of upwelled water further to the north (Fig. 60), resulting in no distinct increase of productivity (Hughes and Barton, 1974; Schemainda et al., 1975; Shaffer, 1976; Tomczak and Hughes, 1980). South of about $15^{\circ}N$ the climate is more humid and fluvial nutrient supply becomes a more important factor in oceanic productivity (Schemainda et al., 1975; Diester-Haass, 1983).

The modern upwelling and high-productivity system is well documented in the surface sediments by several signals such as high contents of organic carbon and biogenic opal, stable isotope records, and foraminifera assemblages (Koopmann, 1981; Ganssen and Sarnthein, 1983; Thiede, 1983; Tiedemann, 1985). From detailed investigations of piston cores from the Northeast Atlantic, informations about changes in paleoproductivity and paleoclimate during the Pleistocene are also available (Fig. 61; Müller and Suess, 1979; Müller et al., 1983; Sarnthein et al., 1988; Mix, 1989). Evidence from organic carbon and foraminifera transfer functions suggest that in general productivity was higher during glacial intervals than today, especially in upwelling regimes. The results obtained by different techniques disagree, however, in detail (Mix, 1989). Furthermore, nutrient supply by rivers which was increased during the more humid interglacials, may have locally caused increased productivity, too (Diester-Haass, 1983; Stein, 1985a).

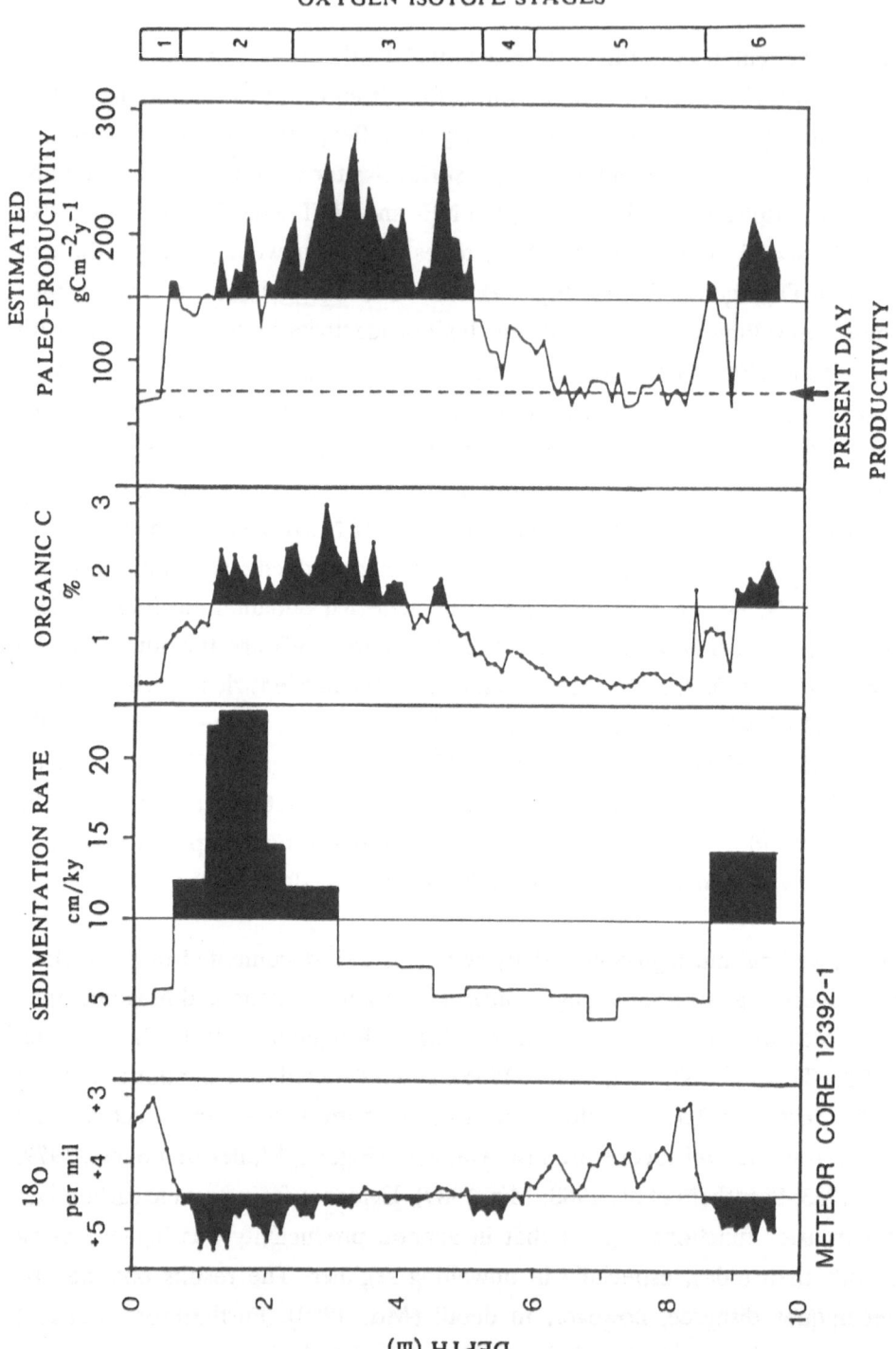

Fig. 61: δ^{18}O values, sedimentation rates, organic carbon contents, and paleoproductivity estimates at "Meteor"-Core 12392 from the continental rise off Spanish Sahara, Northwest Africa (from Müller and Suess, 1979 and Müller et al., 1983).

The influences of both upwelling intensity and fluvial nutrient supply on oceanic productivity can be studied in the sediments from Site 658 (Ruddiman, Sarnthein et al., 1988; Stein et al., 1989b; Tiedemann, et al., 1989).

Deep-water circulation

The main two deep-water masses in the Northeast Atlantic are the well-oxygenated North Atlantic Deep Water (NADW) and the Antarctic Bottom Water (AABW) (Fig. 60). In the deep basins of the Northeast Atlantic, the AABW is a mixture of NADW and AABW masses. The upper boundary of this mixed bottom water lies near 4350m in the Sierra-Leone Rise area and gradually rises northward to about 3900m south of the Canary Islands (Fig. 60; Hobart et al., 1975; Lonsdale, 1978; Ruddiman, Sarnthein et al., 1988). During glacial maximum, reduced production of NADW may have caused a shallower depth of this boundary and a lower dissolved oxygen concentration in the deep basins which may have affected organic carbon preservation (Curry and Lohmann, 1983; Curry et al., 1988). These glacial/interglacial changes could also have influenced organic carbon deposition at Sites 660 and 661 close to the boundary between NADW and AABW (Fig. 60).

6.3. Geological setting and sediments

Site 657 is located on the lower continental rise 380 km west of Cap Blanc, outside the present-day influence of coastal upwelling (Fig. 57, Tab. 5). The 178m thick sediment sequence of Miocene to Holocene age can be divided into two lithologic units. The Pliocene to Pleistocene interval consists of light-colored nannofossil ooze alternating with darker colored greenish-gray, silt-bearing ooze (Fig. 62). Two distinct slumps and several turbiditic sequences are intercalated (Ruddiman, Sarnthein et al., 1988; Faugères et al., 1989a). The late Miocene interval, separated from the overlying interval by a hiatus of about 1.5 my (Fig. 63), is dominated by brownish nannofossil bearing and barren clay/silty clay (Fig. 62). The sedimentation rates at Site 657 vary between 20 and 45 m/my in the Pliocene/Pleistocene and between 2.5 and 7 m/my in the Miocene (Fig. 63; Ruddiman, Sarnthein et al., 1988).

Site 658 was occupied on the continental slope 160 km west of Cap Blanc, underneath a major coastal upwelling cell (Fig. 57, Tab. 5). The 300 m thick sequence of lower Pliocene to Holocene hemipelagic sediments is characterized by cyclic changes between light nannofossil ooze and olive to gray calcareous, siliceous, and siliciclastic sediments (Fig. 62).

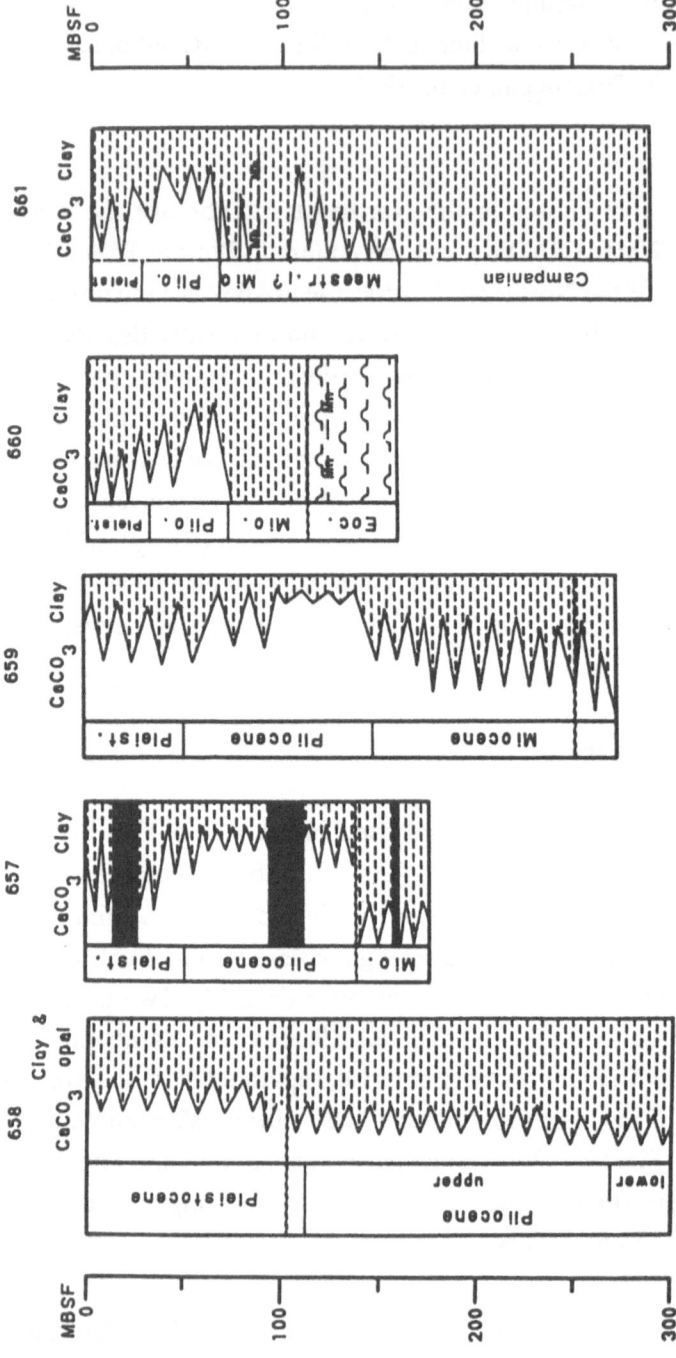

Fig. 62: Scheme indicating major lithologies at Sites 657 - 661 (from Ruddiman, Sarnthein, 1988 and Stein and Faugères, 1989). "Mn" = manganese oxid layers.

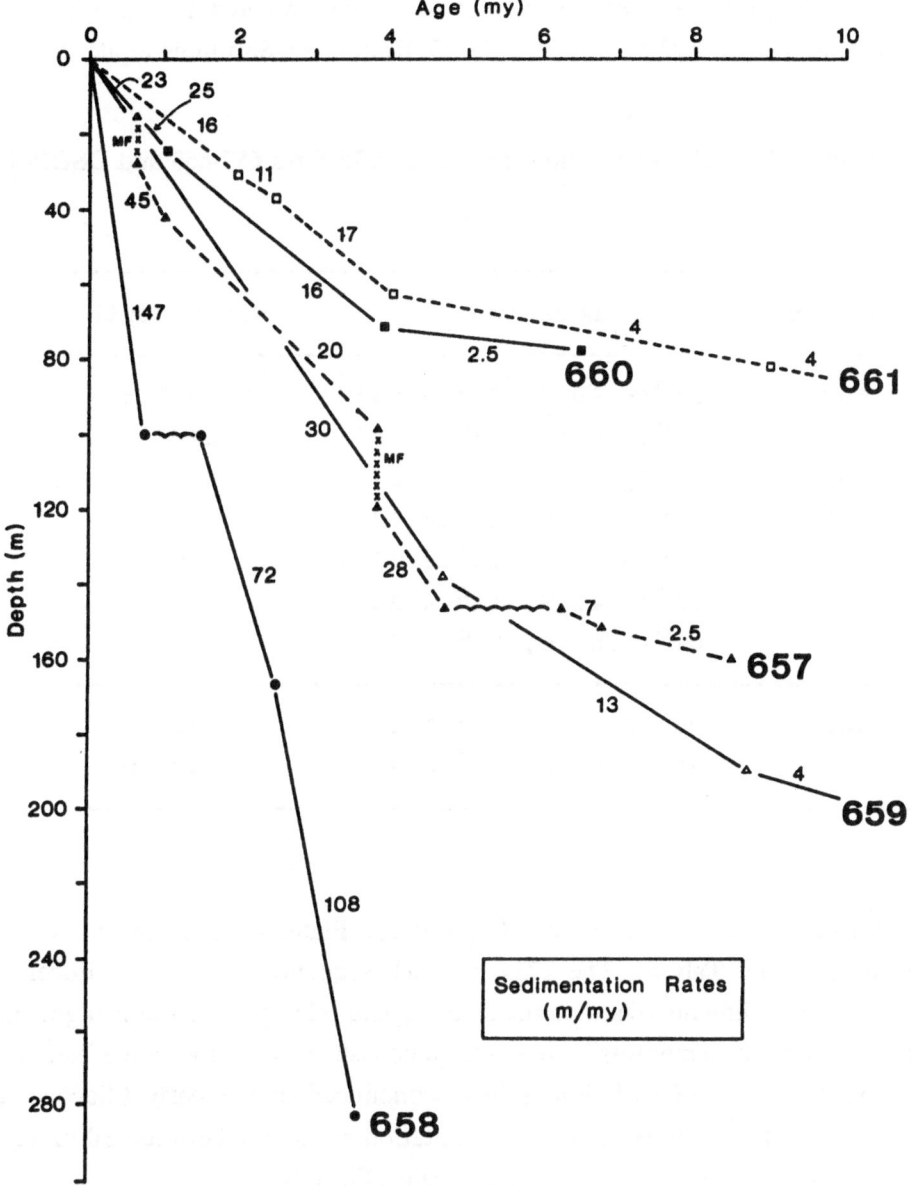

Fig. 63: Age-depth-diagrams and average sedimentation rates at Sites 657 - 661 (after Ruddiman, Sarnthein, et al., 1988). "MF" = mud flow.

Mean sedimentation rates vary between 42 and 147 m/my. In about 100 mbsf, a distinct hiatus lasting from 1.57 to 0.73 Ma occurs (Fig. 63; Ruddiman, Sarnthein et al., 1988).

Tab. 5: Geographical locations and water depths of ODP-Sites 657-663 and DSDP-Sites 366 and 397

ODP-Site	Location	Water depth
657	$21^{o}19.89'$N, $20^{o}56.93'$W	4227 m
658	$20^{o}44.95'$N, $18^{o}34.85'$W	2263 m
659	$18^{o}04.63'$N, $21^{o}01.57'$W	3070 m
660	$10^{o}00.81'$N, $19^{o}14.74'$W	4328 m
661	$09^{o}26.81'$N, $19^{o}23.17'$W	4006 m
662	$01^{o}23.41'$S, $11^{o}44.35'$W	3821 m
663	$01^{o}11.87'$S, $11^{o}52.71'$W	3706 m
366	$05^{o}40.70'$N, $19^{o}51.10'$W	2853 m
397	$26^{o}50.70'$N, $15^{o}10.80'$W	2900 m

Site 659 is located on top of the smooth Cape Verde Plateau, northeast of the Cape Verde Islands (Fig. 57, Tab. 5). The 273.8 m thick sequence consists of Miocene to Holocene pelagic sediments characterized by cyclic changes between light gray nannofossil/ foraminifer-nannofossil ooze and yellowish brown silty nannofossil ooze (Fig. 62). The amplitude of variation is less pronounced in the early Pliocene and strongest in the middle/early Miocene. Sedimentation rates vary between 30 m/my in the Pliocene-Pleistocene and 4 m/my in the Miocene (Fig. 63).

Sites 660 and 661 (Fig. 57, Tab. 5) are located on the Sierra Leone Rise, close to the Kane Gap. The Kane Gap separates the rise into two parts: the Guinea continental margin in the east and the outer Sierra Leone Rise in the west. The gap is a major passage for deep-water exchange between the southern Sierra Leone Abyssal Plain and the northern Gambia Abyssal Plain at least since Miocene times, possibly since the end of Eocene times (Jacobi and Hayes, 1982; Mienert, 1986). Thus, in the Kane Gap area, Tertiary sedimentary processes are strongly influenced by bottom-current activities, which have caused erosion and/or non-deposition at Sites 660 and 661 (Ruddiman, Sarnthein et al., 1988; Stein and Faugères, 1989).

At Site 660, near the northern end of the Kane Gap, a 164.9 m thick sediment sequence of Eocene to Holocene age was recovered (Fig. 62) which can be divided into three different units. The upper 75 m consist of uppermost Miocene to Holocene sediments characterized by cycles of dark gray silty clay and light olive gray (muddy) nannofossil ooze. The interval between 75 and 115.8 mbsf is of Miocene age and is composed of yellowish brown clay. The lower 49.1 m, separated from the overlying sediments by a major hiatus (Fig. 62), consist of middle Eocene, massive, relatively coarse-grained yellowish radiolarian ooze with common pale brown to dark gray laminations (manganese oxide layers; Wosnitza, pers. comm., 1988) (Fig. 62; Ruddiman, Sarnthein et al., 1988). Mean sedimentation rates vary between 25 and 16 m/my during Plio-Pleistocene times, decreasing to 2.5 m/my during late Miocene times (Fig. 62).

At Site 661, located closest to the Kane Gap, a 296.1 m thick sequence of Campanian to Holocene sediments was drilled (Fig. 62). The upper 72.6 m contain uppermost Miocene to Holocene cyclic gray foraminifer-nannofossil ooze to gray muddy nannofossil ooze or clay. The interval between 72.6 and 90.8 mbsf consists of lower upper Miocene brownish to reddish clay. Below 90.8 m, the sediment is composed of Campanian to Tertiary bluish-green, partly zeolite-rich clay and claystone with interbedded Maastrichtian nannofossil oozes in the middle part. Two horizons with striking color changes, metallic nodules, and erosional features underlay the top of this interval (Ruddiman, Sarnthein et al., 1988; Stein and Faugères, 1989). Sedimentation rates vary between 11 and 17 m/my in the uppermost Miocene to Holocene and decrease to 4 m/my in the late Miocene (Fig. 63).

Site 662 was occupied in the eastern equatorial Atlantic (Fig. 57, Tab. 5), underneath the Equatorial Divergence Zone, on the upper eastern flank of the Mid-Atlantic Ridge just south of the Romanche Fracture Zone. At this site, a 200 m thick sediment sequence composed of nannofossil and foraminifera-nannofossil oozes of late Pliocene through Holocene age, was recovered (Ruddiman, Sarnthein et al., 1988). Secondary components include clay, diatoms, and radiolarians. The amplitude of variations between more carbonate-rich facies and more clay-rich facies increased near 2.5 Ma. In the Pleistocene section, major slumps, debris flows, and turbidites are intercalated. Sedimentation rates of the pelagic sediments average 42 m/my (Ruddiman, Sarnthein et al., 1988).

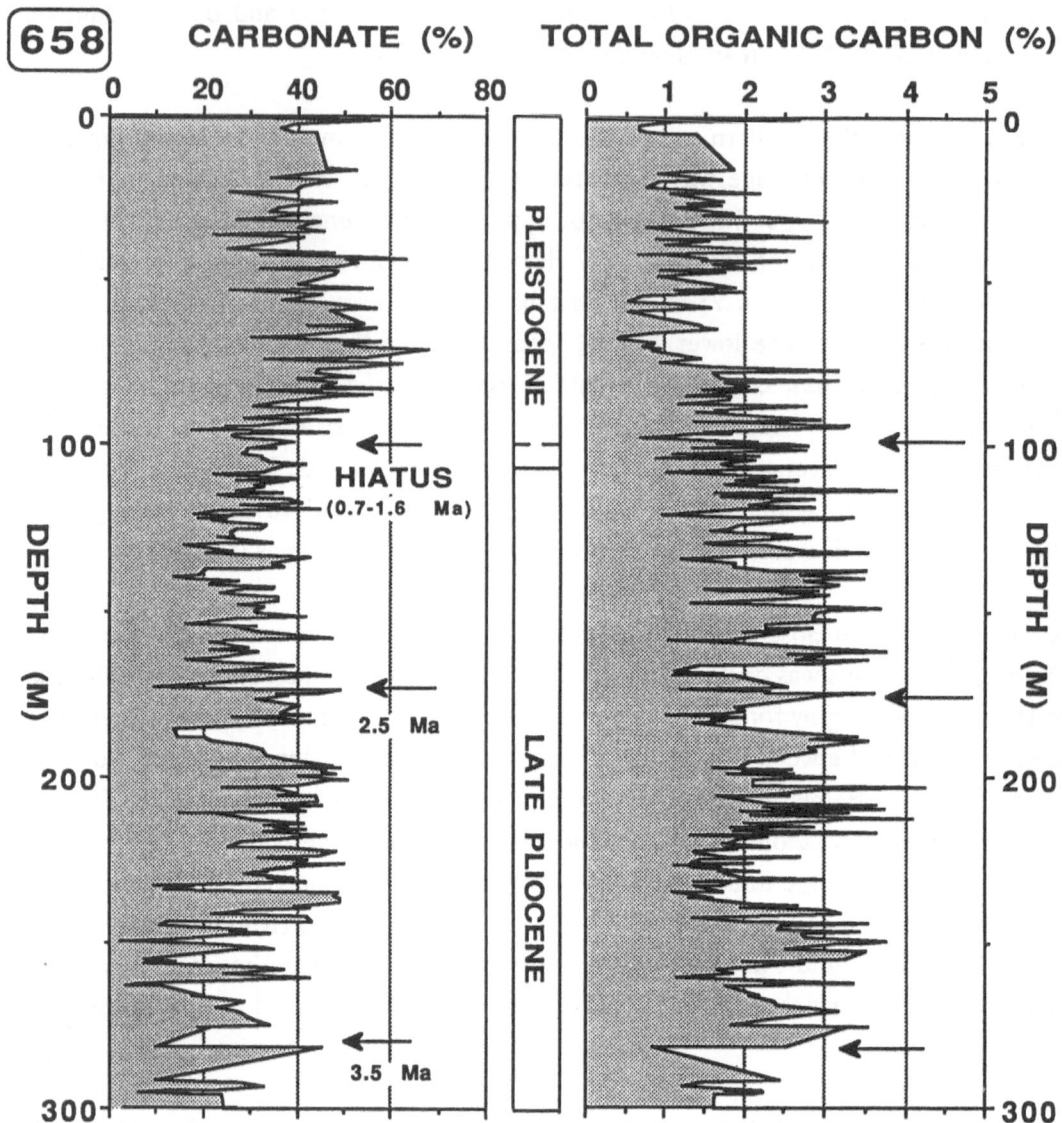

Fig. 64: Carbonate and total organic carbon contents vs. depth at Site 658.

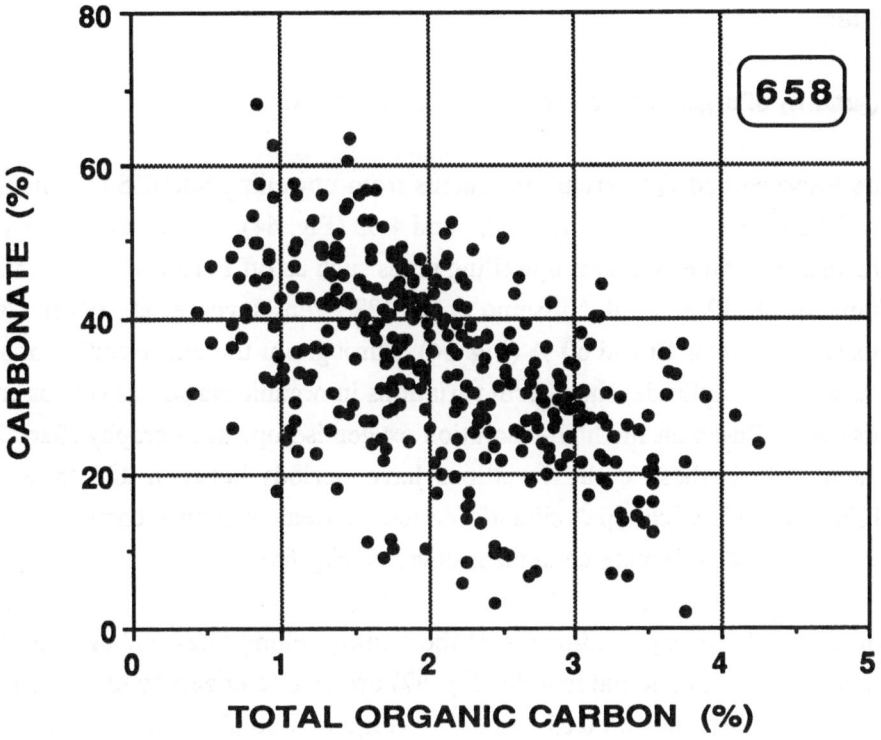

Fig. 65: Carbonate vs. total organic carbon contents at Site 658.

6.4. Results

6.4.1. Quantity of organic carbon and carbonate content

The late Pliocene and Quaternary sediments from upwelling-Site 658 are characterized by high TOC values ranging between 0.5 and 4 % (Fig. 64). These values are similar to those recorded in other modern upwelling areas such as off Peru and off Arabia (Suess, von Huene et al., 1988; Prell, Niitsumo et al., 1988; ten Haven et al., 1990). Carbonate contents vary between 10 and 60 % (Fig. 64). Throughout the entire sediment sequence of Site 658, high-amplitude, short-term variations in organic carbon as well as carbonate contents occur. Based on the high-resolution oxygen isotope stratigraphy (Sarnthein and Tiedemann, 1989), these cyclic variations have periods between 100 ky and 20 ky, resembling "Milankovitch-type" climatic cycles. A weak negative correlation between organic carbon and carbonate contents is obvious (Fig. 65).

In comparison, the pelagic sediments of the non-upwelling Sites 657 and 659 from the Northwest African continental margin (Fig. 57) are characterized by low organic carbon contents with values between 0.05 and 0.5 % (Figs. 66 and 67). These values are typical of modern open-ocean environments. At Site 657, samples with high organic carbon contents of 1 to 3 % are restricted to turbidites and slumps (Fig. 66). Carbonate contents are generally higher in the pelagic sediments at Sites 657 and 659 than in the hemipelagic sediments at Site 658 (Figs. 66 and 67). The carbonate content of the turbidites at Site 657, on the other hand, is also relatively low (20 to 45 %; Fig. 66).

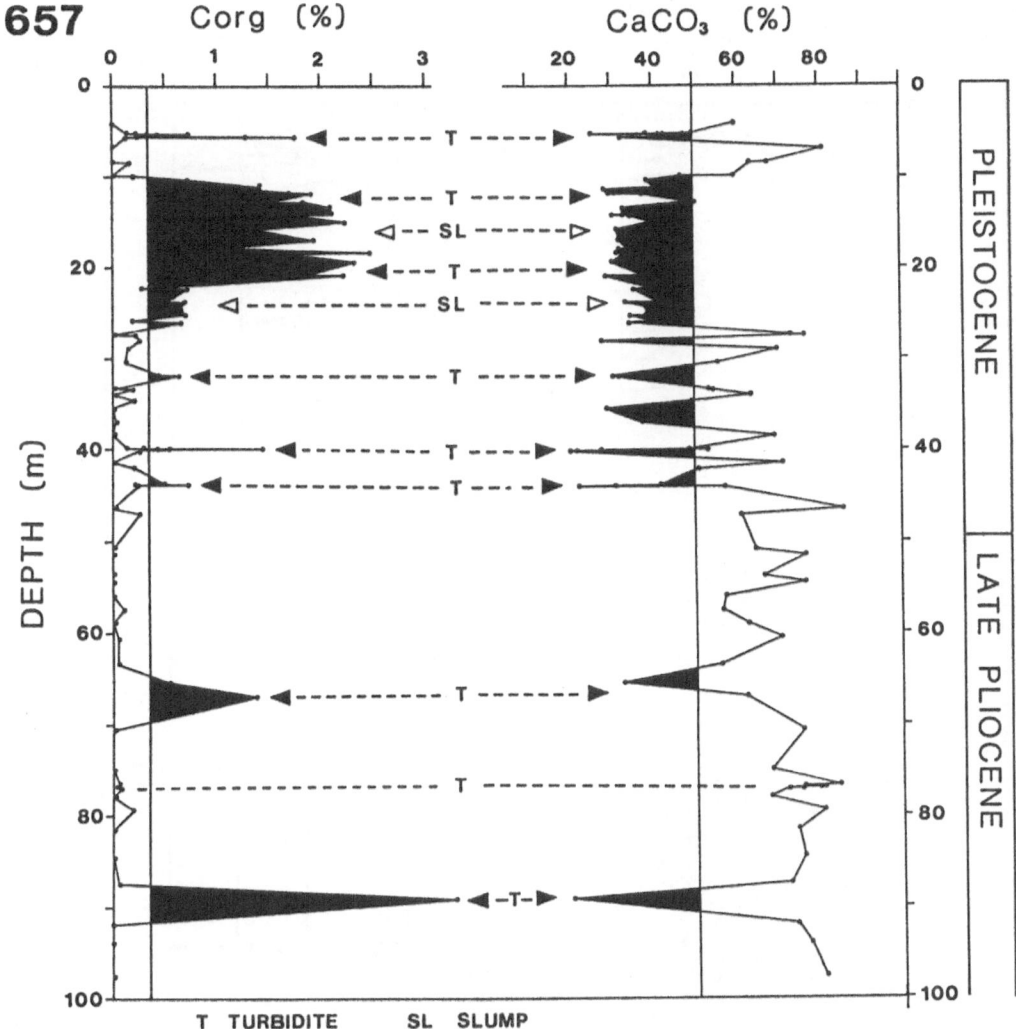

Fig. 66: Carbonate and total organic carbon contents at Site 657. Black triangles mark turbidites (T), open triangles mark slumps (SL).

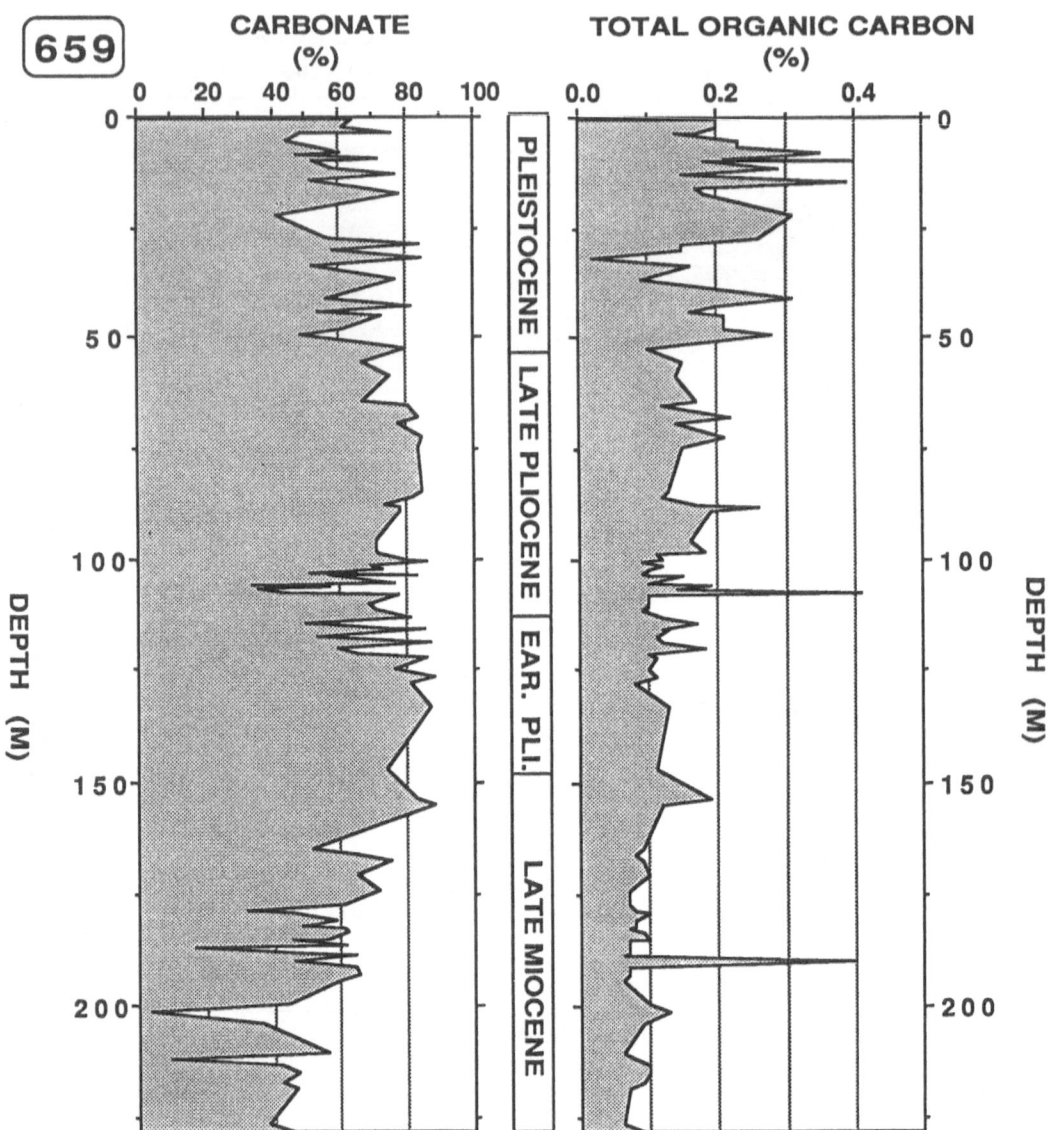

Fig. 67: Carbonate and total organic carbon contents vs. depth at Site 659.

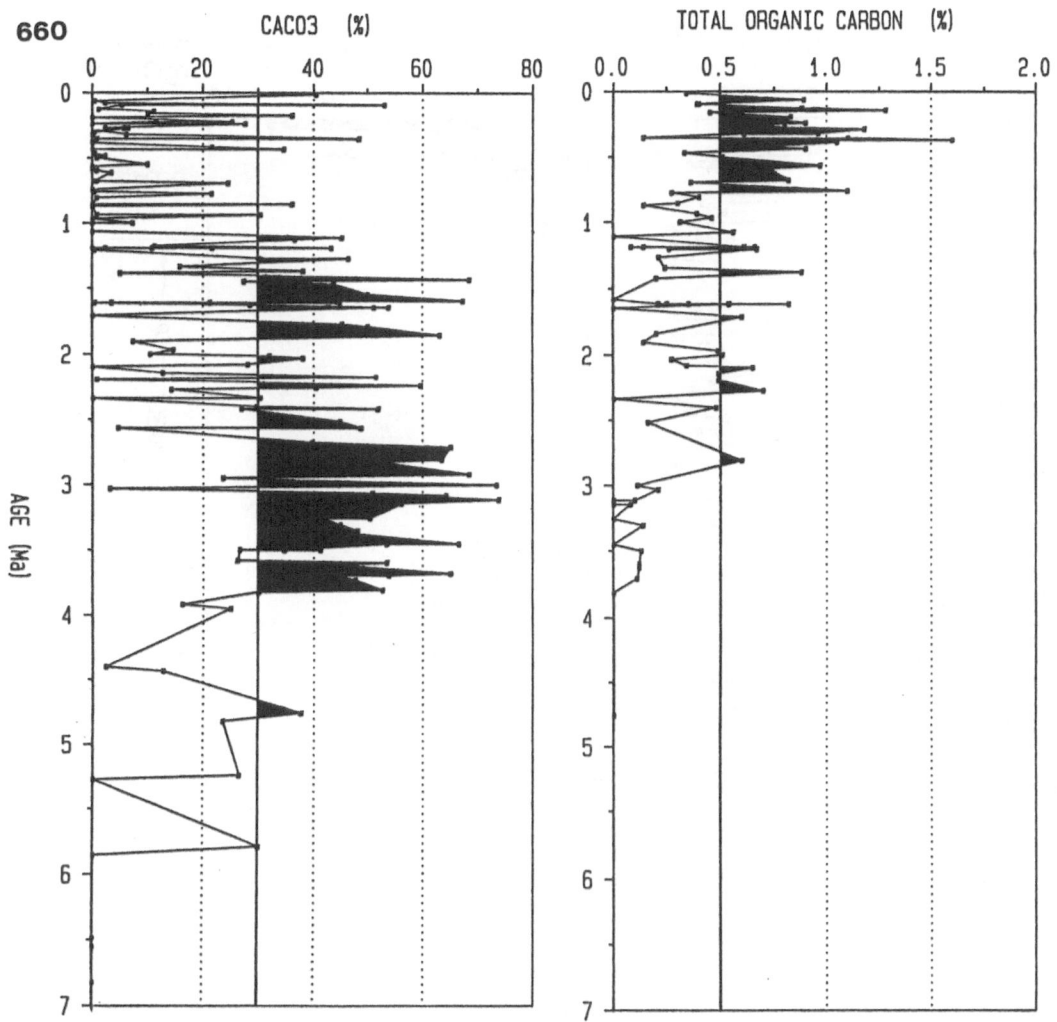

Fig. 68: Carbonate and total organic carbon contents vs. depth at Site 660.

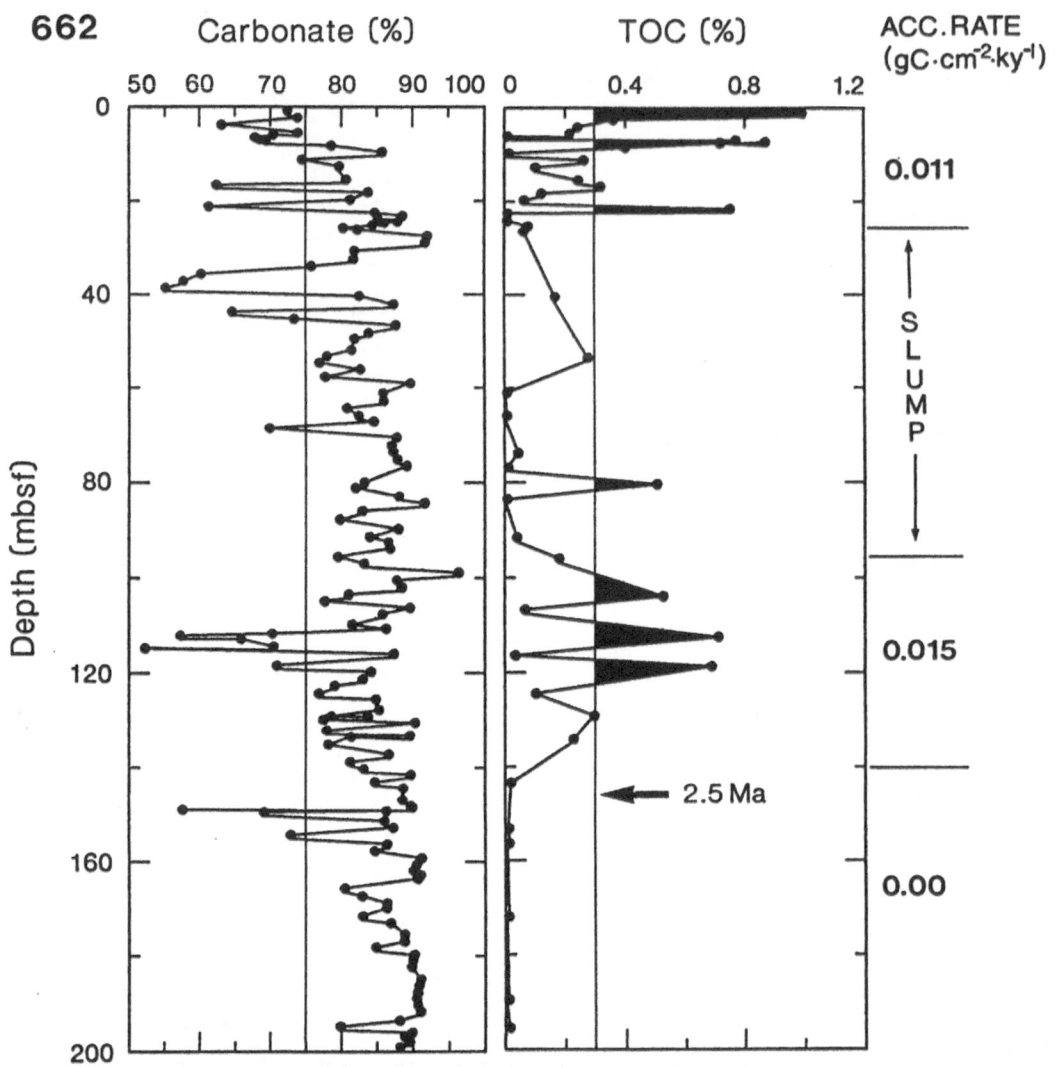

Fig. 69: Carbonate and total organic carbon contents vs. depth and mean organic carbon accumulation rates at Site 662 (after Ruddiman, Sarnthein et al., 1988).

The Plio-Pleistocene sediments at Site 660 are influenced by the North Equatorial Divergence Zone and show total organic carbon contents between 0 and 1.5 % (Fig. 68). The sequence can be divided into three intervals based on distinct changes in organic carbon content near 2.8 (to 2.5) and 0.75 Ma: Below 54 mbsf (older than 2.8 Ma), TOC values are less than 0.2 %; between 54 and 19 mbsf (2.8 to 0.75 Ma), TOC values vary between 0 and 0.9 %; and in the upper 19 m (i.e., the last 0.75 Ma), TOC values vary between 0.3 and 1.5 % (Fig. 68). The carbonate contents are characterized by high-amplitude fluctuations between 0 and 70 %, with maximum values occurring between about 70 and 25 mbsf (i.e., 3.8 and 1 Ma) (Fig. 68).

At Equatorial-upwelling Site 662, total organic carbon contents of the upper Pliocene to Pleistocene sediments vary between 0 and 1 % (Fig. 69; Ruddiman, Sarnthein et al., 1988). Based on the relatively small data set available until now, the organic carbon record can be described as follows: Below 135 mbsf (> 2.2 Ma), TOC values are close to zero. Near 135 mbsf, the organic carbon content increases and the upper 135 m (2.2 to 0 Ma) are characterized by values ranging between 0 and 1 %, with the lower values concentrating in the slump interval between 26 and 100 mbsf. Carbonate contents vary between 60 and 90 % (Fig. 69).

As outlined in the previous chapters, changes in percentage values of organic carbon can result from changes in both organic carbon supply and sedimentation of mineral matter (mainly siliciclastics and carbonates). Thus, organic carbon percentages of sediments from Sites 658, 659, 660, and 662 were transformed into accumulation rates.

At Sites 658, mean accumulation rates of organic carbon vary between 0.08 and 0.5 gC cm^{-2} ky^{-1}, with increasing amplitude of fluctuations during the last 0.7 Ma (i.e., above the hiatus) (Fig. 70).

At Site 659, mass accumulation rates of organic carbon are 1-2 orders of magitude lower than those at Site 658 (Fig. 71). They vary between 0.003 and 0.01 gC cm^{-2} ky^{-1} during the last 4.5 Ma. In the late Miocene, extremely low accumulation rates of less than 0.002 gC cm^{-2} ky^{-1} were recorded.

At Site 660, accumulation rates of organic carbon vary between 0 and 0.02 gC cm^{-2} ky^{-1} during the last 5 Ma (Fig. 71): Prior to 2.4 Ma, the accumulation rates are below 0.005 gC cm^{-2} ky^{-1}. Near 2.4 Ma, a distinct increase to 0.01 gC cm^{-2} ky^{-1} occurs. Maximum accumulation rates of 0.01 to 0.02 gC cm^{-2} ky^{-1} were calculated for the last 0.5 Ma.

At Site 662, a similar increase in accumulation rates of organic carbon from about 0 to 0.015 gC cm^{-2} ky^{-1} also occur near 2.5 Ma (Fig. 69).

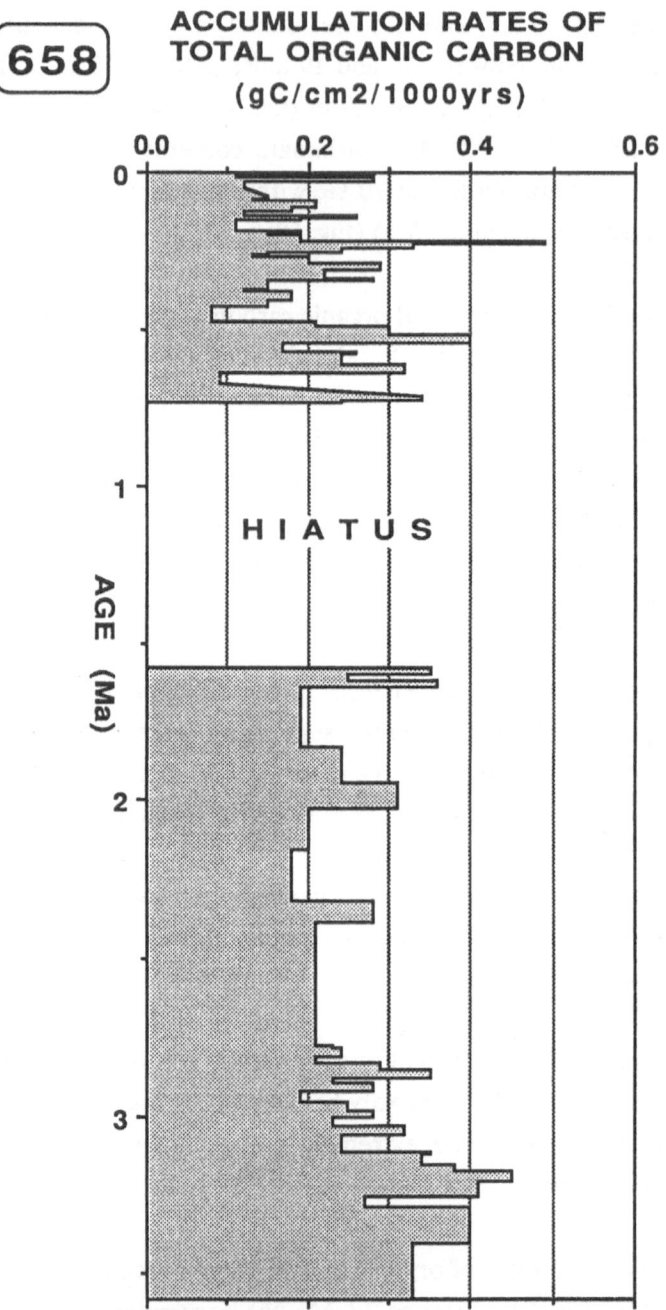

Fig. 70: Mean accumulation rates of total organic carbon at upwelling-Site 658, based on mean sedimentation rates from Sarnthein and Tiedemann (1989).

ACCUMULATION RATES OF ORGANIC CARBON (gC·cm⁻²·ky⁻¹)

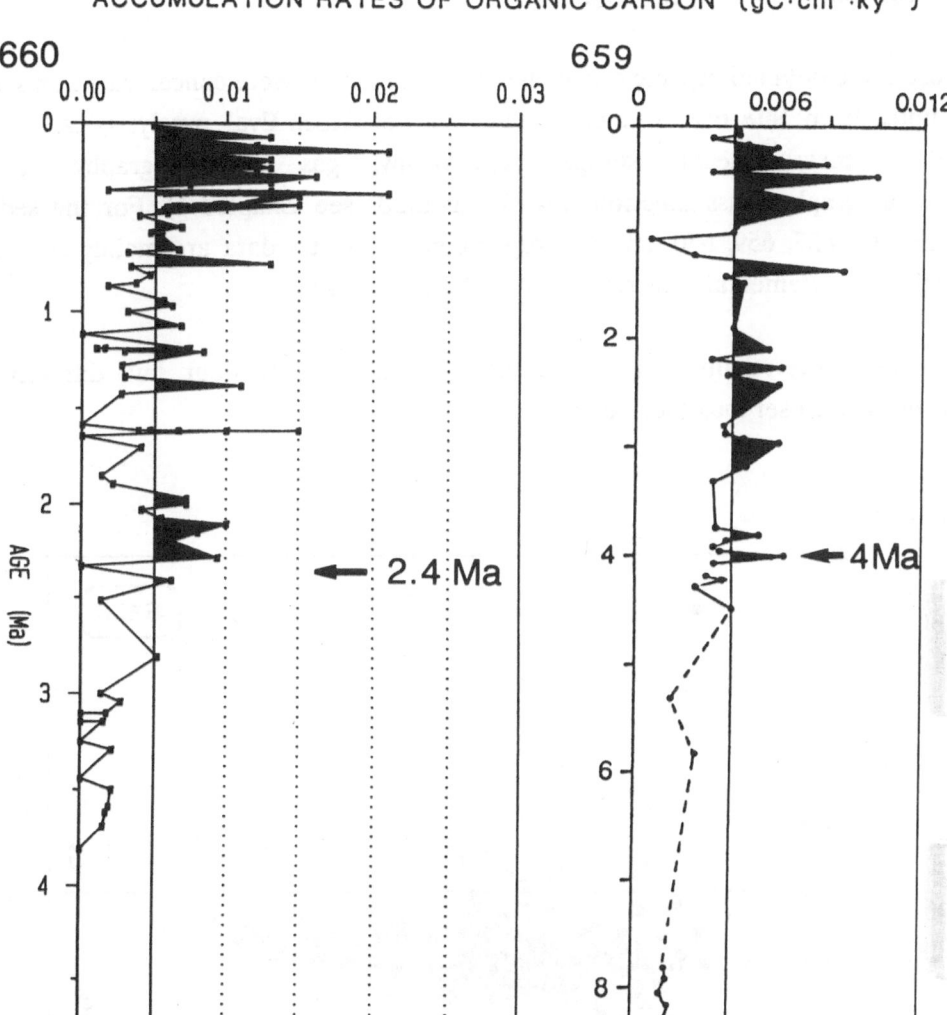

Fig. 71: Accumulation rates of total organic carbon at Sites 659 and 660.

6.4.2. Composition of organic matter

Most of the detailed organic geochemistry investigations were concentrated on sediment samples from Site 658, i.e., elemental analyses, Rock-Eval pyrolysis, carbon stable isotope measurements, kerogen microscopy, gas chromatography, and gas chromatography/mass spectrometry (for methods see Chapter 4). For the sediments from Sites 657, 659, 660, and 662 organic geochemistry data are mainly restricted to results from elemental analyses and Rock-Eval pyrolysis.

In general, the results of the different methods are consistent, but, differences are obvious at a closer look (see below).

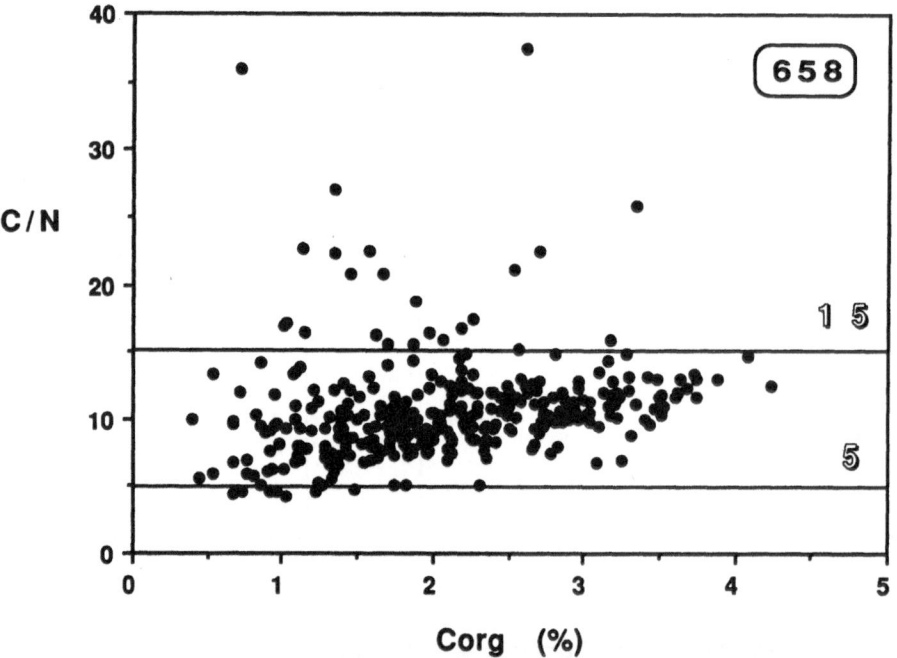

Fig. 72: Total organic carbon contents vs. C/N ratios at Site 658.

Organic carbon/nitrogen (C/N) ratios

C/N ratios between 5 and 15 are common in the sediments at Site 658 (Fig. 72), suggesting a mixed marine/terrigenous type of organic matter with a dominance of the marine proportion. However, in several samples higher C/N ratios of 20 to 35 were also determined (Fig. 72) which may indicate a higher proportion of terrigenous organic matter in these samples. In general, there is no correlation between C/N ratios and organic carbon contents (Fig. 72).

At Site 660, C/N ratios are similar to those determined at Site 658 (between 5 and 15; Fig. 73), due to a dominance of marine organic matter in these sediments. In general, there seems to be a positive correlation between C/N ratios and TOC values. Higher C/N ratios around 15 coincide with high TOC values and may indicate elevated amounts of terrigenous organic matter. However, this positive correlation should not be overinterpreted because of difficulties in interpretation of C/N ratios from organic-carbon-poor samples (see above).

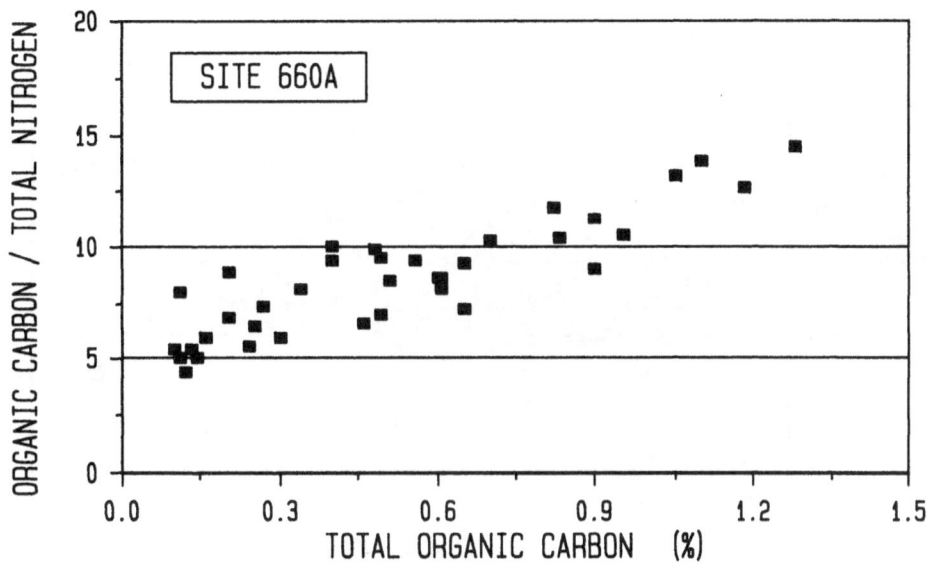

Fig. 73: Total organic carbon contents vs. C/N ratios at Site 660.

Rock-Eval pyrolysis

Hydrogen index values derived from Rock-Eval pyrolysis are relatively high throughout the sediment sequence of Site 658. Most of the values vary between 200 and 400 mgHC/gC (Figs. 74 and 75), suggesting that most of the organic matter in the sediments from Site 658 is a mixture of kerogen types II and III, i.e., a mixture of marine and terrigenous organic material, with a dominance of the marine proportion. Based on the correlation between hydrogen index values and maceral composition for DSDP sediments (Stein et al., 1986), the amount of marine organic matter in Site 658 sediments were estimated (from hydrogen index values) to vary between 40 and 80 % of the organic material (Fig. 23). Rock-Eval pyrolysis and elemental analyses, both performed on kerogen concentrates, support the dominance of marine organic matter: Hydrogen index values vary between 230 and 640 mgHC/gC (Fig. 76) and atomic H/C ratios range from 1.0 to 1.5 (Fig. 77).

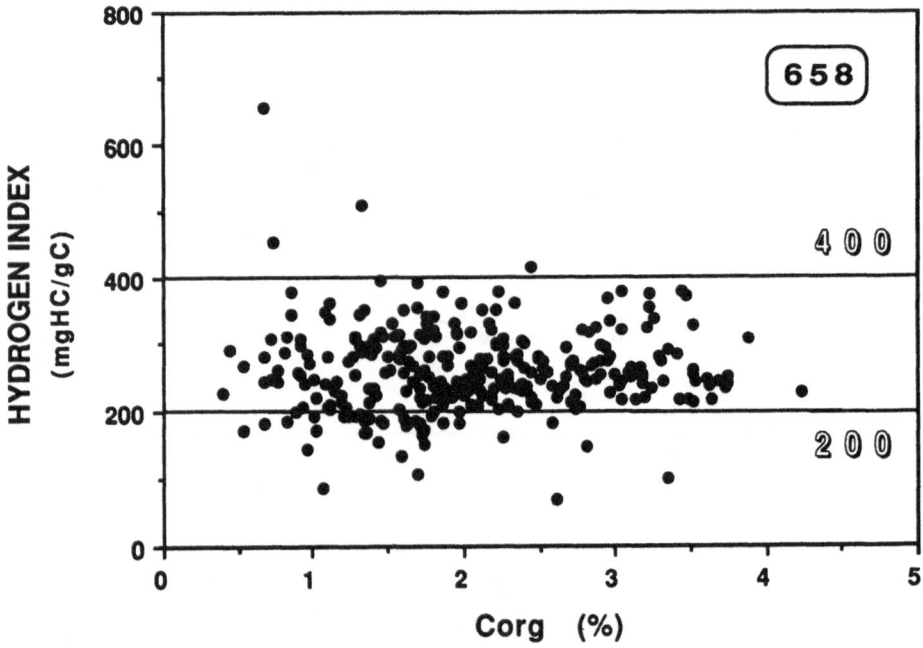

Fig. 74: Total organic carbon contents vs. hydrogen indices at Site 658.

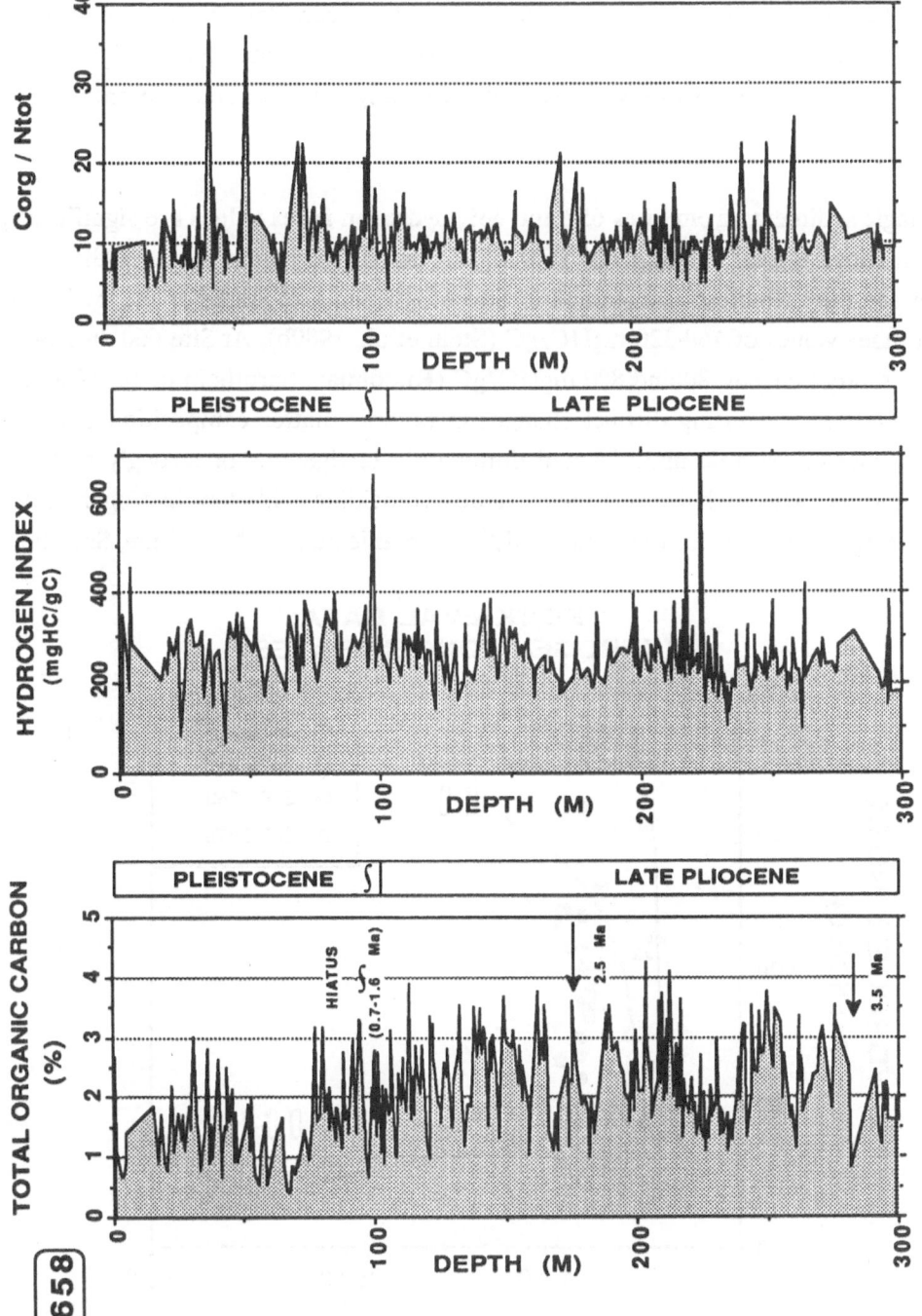

Fig. 75: Total organic carbon contents, hydrogen indices, and C/N ratios vs. depth at Site 658.

Although the records of relative concentrations and accumulation rates of organic carbon at Site 658 display distinct long-term changes during the last 3.5 Ma (Figs. 64 and 70), no such changes are obvious in the records of hydrogen values and C/N ratios (Fig. 75).

In the pelagic sediments from Sites 657 and 659, hydrogen index values are significantly lower than those typical for Site 658 (50 to 120 mgHC/gC; Stein et al., 1989b). An exception are the turbidites and slumps at Site 657 which are characterized by higher hydrogen index values of 160-320 mgHC/gC (Stein et al., 1989b). At Site 660, hydrogen index values vary between 80 and 800 mgHC/gC (Ruddiman, Sarnthein et al., 1988; cf., Figs. 99 and 100), suggesting distinct changes in organic matter composition between dominantly marine organic material and dominantly terrigenous or strongly oxidized organic material. The high oxygen index values coinciding with low hydrogen index values indicate the occurrence of strongly oxidized organic matter (Ruddiman, Sarnthein et al., 1988).

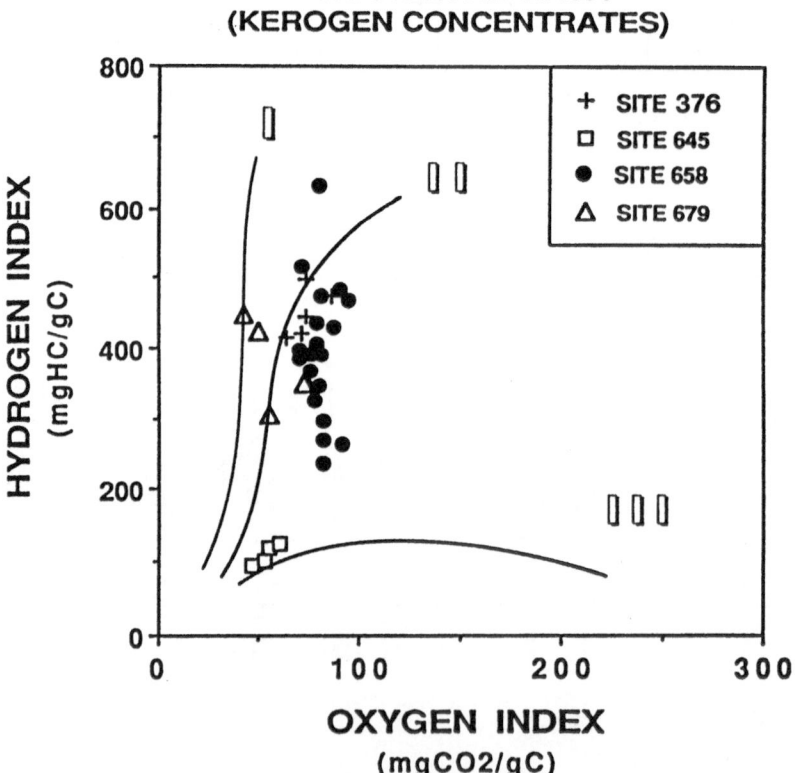

Fig. 76: Rock-Eval pyrolysis analyses performed on kerogen concentrates from sediments of Site 658: Hydrogen- vs. oxygen indices. For comparison also data from Site 645 (Baffin Bay), Site 679 (upwelling off Peru), and Site 376 (Mediterranean sapropels) are shown.

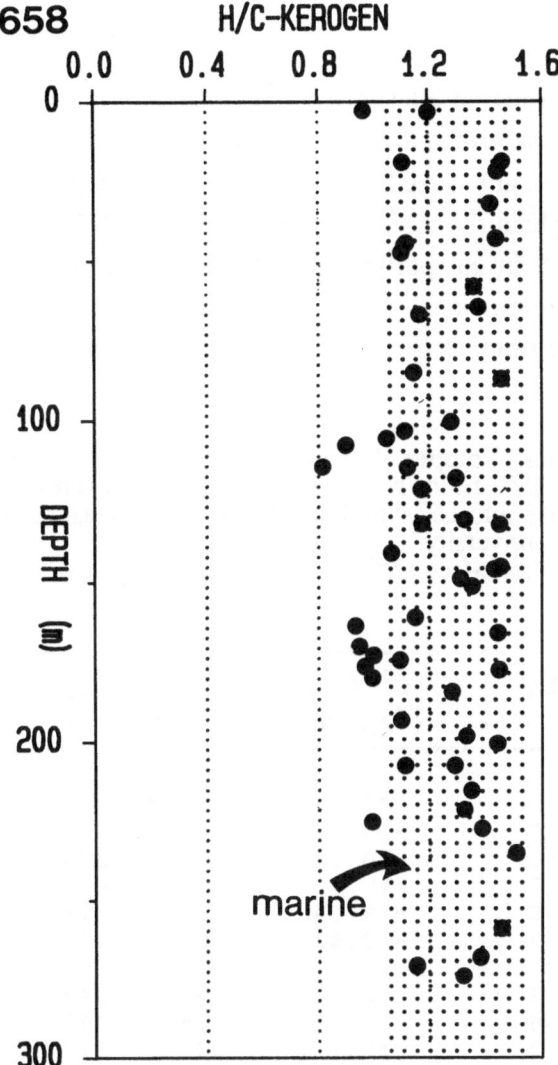

Fig. 77: Atomic H/C determined on kerogen concentrates from sediments of Site 658.

Stable carbon isotopes

For a set of 28 samples from Sites 657 and 658, carbon stable isotopes of the organic matter fraction were determined (Tab. 6). Most of the $\delta^{13}C_{org}$ values of Site 658 sediments vary between about -20.5 and -19.5 $^{O}/_{OO}$ indicating the dominance of marine organic matter throughout the sediment section (Fig. 78; cf. Chapter 4.2, Fig. 26). It has to be considered that, in comparison to C/N ratios and hydrogen indices, $\delta^{13}C_{org}$ values were only determined on a limited number of samples. The $\delta^{13}C_{org}$ values measured on turbidite samples from Site 657 are very similar to those at Site 658 (Tab. 6).

Tab. 6: Stable carbon isotopes, hydrogen index (HI bulk sediment; HI_k kerogen concentrates), atomic H/C ratios measured on kerogen concentrates, and C/N ratios of sediments from Site 658 and 657. Stable carbon isotope measurements were performed by P. Müller, Bremen University.
* = measured on carbonate-free sediment samples.

Sample	Depth (m)	$\delta^{13}C_{org}$ ($^{O}/_{OO}$PDB)	HI	HI_k (mgHC/gC)	H/C	C/N
658						
03-6-040	23.10	-21.30	86	–	–	13
04-6-040	32.60	-19.60	168	–	1.41	12
08-2-040	64.60	-20.23	297	263	1.36	11
10-3-040	85.10	-19.50	310	469	1.17	11
12-1-040	101.06	-19.54	259	296	1.29	10
13-3-040	112.97	-19.41	308	392	0.82	13
13-4-040	114.24	-19.59	362	407	1.12	9
14-2-040	121.12	-19.59	291	347	1.18	11
15-3-051	132.06	-19.69	328	347	1.17	12
16-3-040	141.89	-19.76	381	390	1.06	11
17-6-046	156.16	-19.87	262	473	–	11
20-3-040	169.56	-20.66	180	482	0.95	14
20-9-034	174.96	-19.59	246	631	1.17	12
23-4-040	199.70	-19.93	214	515	1.33	12
24-2-063	207.14	-19.79	273	396	1.29	8
27-3-120	238.02	-19.68	236	–	–	16
29-2-032	254.62	-19.82	232	–	–	12
30-6 087	270.77	-19.58	239	–	1.49	12
31.2.040	273.80	-19.37	–	392	1.32	–
33-2-040	292.80	-20.11	–	–	–	–
657						
01-4-099	5.49	-21.12	298*	–	–	11
01-4-104	5.54	-20.33	275*	–	–	10
01-4-112	5.62	-19.83	380*	–	–	16
01-4-116	5.66	-20.84	–	–	–	–

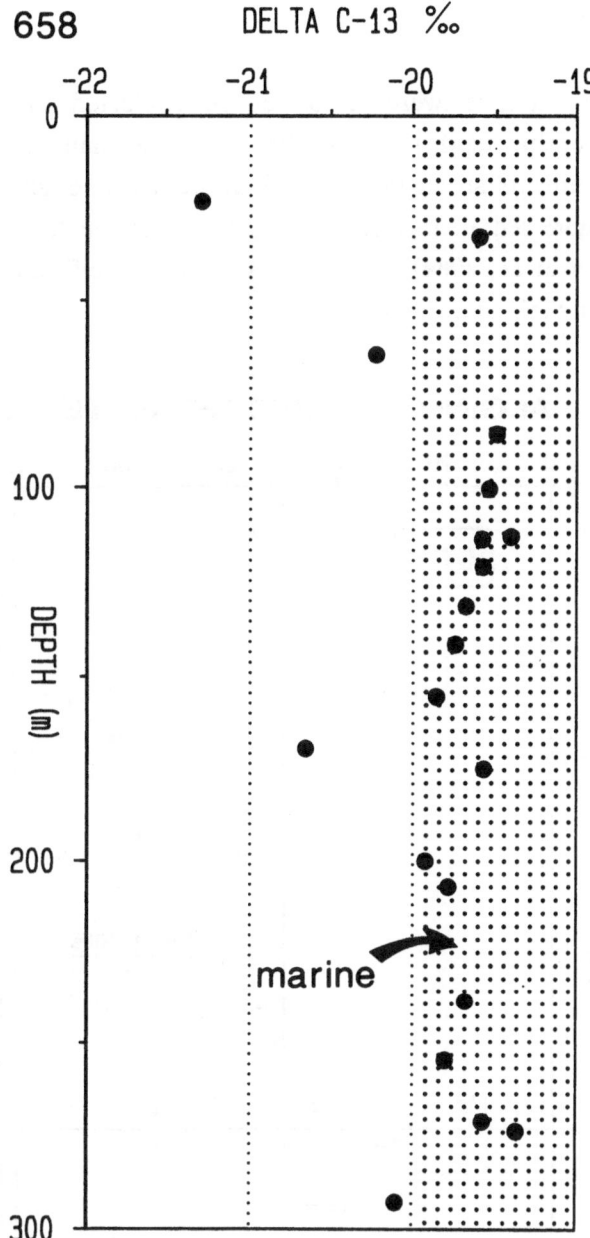

Fig. 78: $\delta^{13}C_{org}$ values of Plio-Pleistocene sediments from Site 658 vs. age ($\delta^{13}C_{org}$ measurments were performed by P. Müller, Bremen University).

Kerogen microscopy

Kerogen microscopy analyses were performed on a selected set of samples from Site 658, covering the entire depth range at that site. The main macerals in the organic matter of these sediments are alginite, liptodetrinite, vitrinite, and inertinite; sporinite, common in near-shore sediments from other locations in the Atlantic (e.g., Cornford et al., 1979; Rullkötter et al., 1984), was not observed in significant amounts in Site 658 sediments (Littke in Stein et al., 1989b).

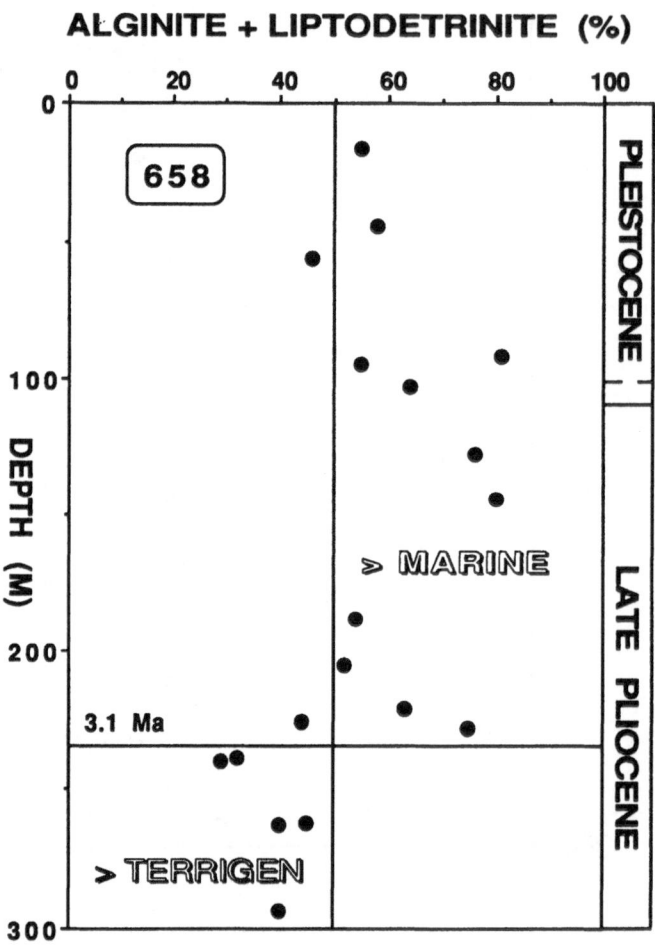

Fig. 79: Amounts of alginites plus liptodetrinites (% of total macerals) vs. depth at Site 658 (data from Littke in Stein et al., 1989b).

Since (marine) alginite is the only liptinite identified in large quantities, the origin of the liptodetrinites is also interpreted as being marine. Thus, the total marine components (alginite plus liptodetrinite) vary between 30 and 80 %, with the lowest values (30 to 40 %) occurring in sediments from the older part of the sequence (Fig. 79). This means, that the terrigenous macerals inertinite and vitrinite may represent 60 to 70 % of the total macerals in the interval between 3.6 and 3.1 Ma. These values are in agreement with the estimates of organic matter composition based on hydrogen index data (cf., Fig. 23).

In general, the terrigenous macerals are dominated by inertinites; vitrinites are of secondary importance (Stein et al., 1989b). The dominance of inertinites may point toward a stronger degradation (oxidation) of the terrigenous plant fragments during transport or the deposition of reworked kerogen particles from older more mature strata.

According to the results of kerogen microscopy, upper Plio-Pleistocene sediments with increased proportions of marine organic material also have elevated total organic carbon contents (Fig. 80). On the other hand, in sediments from the lower part of the sequence (i.e., sediments older than 3.1 Ma) maximum organic carbon contents coincide with reduced proportions of marine macerals (i.e., maximum proportions of terrigenous macerals).

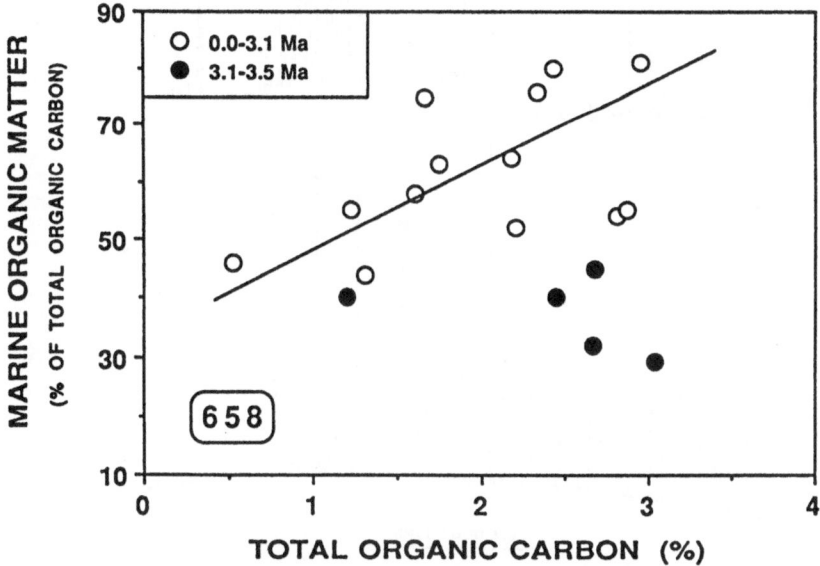

Fig. 80: Proportion of marine organic matter (based on maceral composition) vs. total organic carbon content at Site 658.

Gas chromatography and gas chromatography/mass spectrometry

On a selected set of samples from Sites 658 and 659, gas chromatography and gas chromatograpy/mass spectrometry analyses were preformed (ten Haven et al., 1989). As an example, a gas chromatogram of the "total lipid" fraction and a gas chromatogram of the aliphatic hydrocarbon fraction of Sample 658A-21-6, 7-14cm is shown in Figure 81. Long straight-chain alkanes of terrigenous origin (C_{27}-C_{33}) are the most abundant compounds in the aliphatic hydrocarbon fraction, but they are of minor importance in the total lipid fraction. Marine-derived long-chain unsaturated ketones and steroids are always the most important classes of compounds in the total lipid fraction of samples from Site 658 (for details of compound identification, see ten Haven et al., 1989). These results support the bulk organic geochemistry and kerogen microscopy data, i.e., the dominance of marine organic matter in the Site 658 sediments.

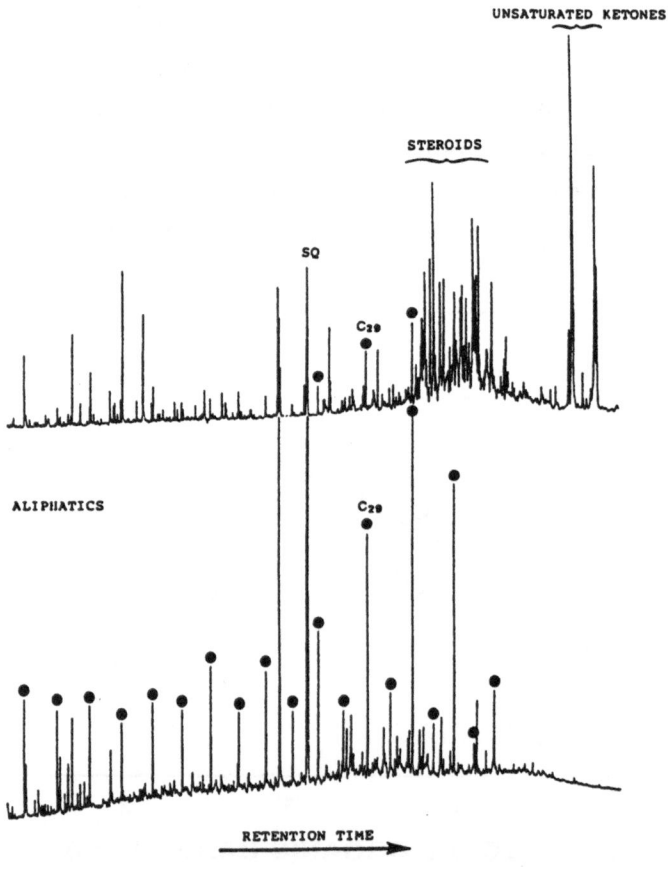

Fig. 81: Gas chromatogram of "total lipid" and of aliphatic fraction from Sample 108-658A-21-6, 7-14 cm (183 mbsf). n-Alkanes are marked by dots, starting with C_{17}. SQ = squalane; internal standard. The "total lipid" extract was derivatized before GC measurements (see ten Haven et al., 1989).

6.4.3. Maturity of organic matter

The low temperatures of maximum pyrolysis yield (T_{max}) ranging between 400 and 430°C, and vitrinite reflectance values between 0.23 and 0.27 % indicate a low maturity of organic matter at Site 658 (Fig. 82). The low abundance of sterenes/steranes in the hydrocarbon fraction and the high abundance of free extractable sterols also suggest that major diagenetic alterations of the organic matter have not occurred (ten Haven et al., 1989).

A second maximum of vitrinite reflectance values in some samples (around 0.5 %; Fig. 82) may imply the presence of redeposited organic matter.

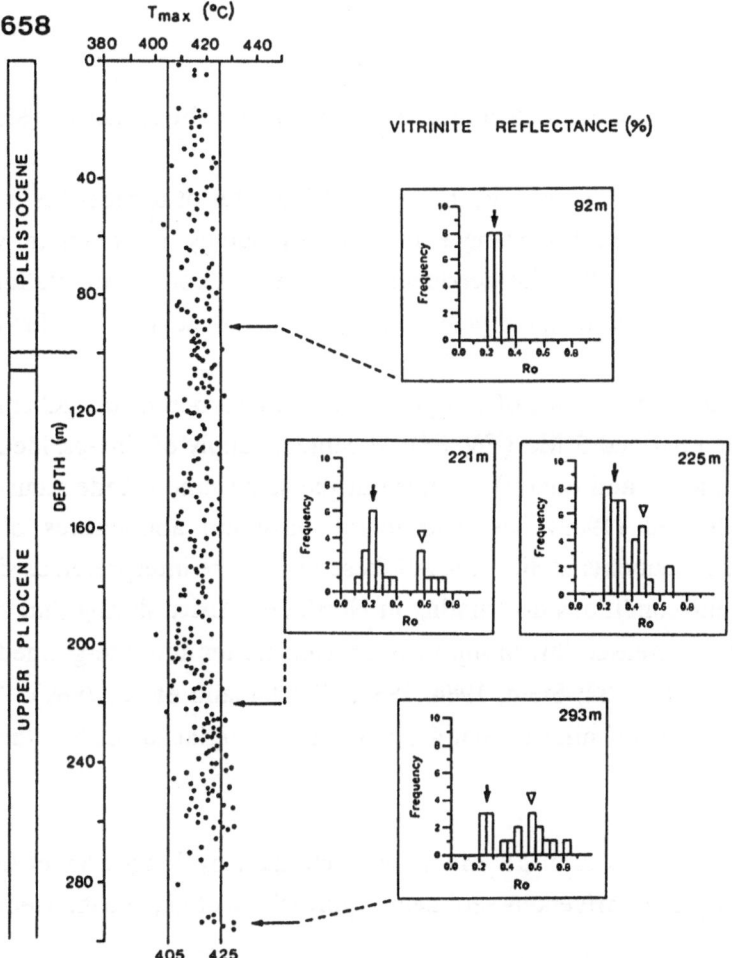

Fig. 82: Temperatures of maximum pyrolysis yield (T_{max}) from Rock-Eval measurements on sediments from Site 658. Vitrinite reflectance histograms are shown for four selected samples on the right-hand side (from Stein et al., 1989b).

6.5. Discussion

The detailed study of organofacies types and accumulation rates of terrigenous and marine organic carbon of Leg-108 sediments may provide important information on the evolution of both the paleoclimate in Northwest Africa and the paleoceanic circulation in the Northeast Atlantic during late Cenozoic times. These results are also important for the calculations of global productivity changes and CO_2 budgets. In addition, evidence for sediment redeposition processes along the Northwest African continental margin is available from the organic carbon data. Last but not least, the results of this study are of interest for economic aspects because the organic carbon-enriched sediments from the upwelling area off Northwest Africa have a high petroleum potential.

6.5.1. Terrigenous organic matter and paleoclimate in Northwest Africa (Site 658)

As outlined in Chapter 2.3, changes in accumulation rates of terrigenous organic matter are primarily controlled by changes in rates of supply rather than variations in preservation. Consequently, the terrigenous organic-carbon accumulation rates may reveal information concerning the paleoclimate in Northwest Africa at that time.

High mean accumulation rates of terrigenous organic carbon of 0.2 gC cm^{-2} ky^{-1} were reached between 3.6 and 3 Ma (Fig. 83). Maximum supply of fine-grained siliciclastics (mainly clay minerals and quartz) occurred at the same time (Tiedemann et al., 1989). Furthermore, Stabell (1989) has determined maximum abundances of fresh-water diatoms in these sediments at Site 658. All these results are interpreted as signals caused by humid climatic conditions dominating in Northwest Africa during that time interval, which resulted in increased fluvial supply of organic matter and fine-grained siliciclastics (cf., Sarnthein et al., 1982; Stein, 1984, 1985a; Tiedemann et al., 1989). The climatic situation was probably similar to that suggested for interglacials during the Quaternary (Fig. 84).

The interval between 3 and 2.8 Ma is characterized by long-term changes in both organic and inorganic terrigenous sediment supply (Stein et al., 1989b; Tiedemann et al., 1989).

Near 2.8 Ma, the abundances of fresh-water diatoms also distinctly decreased in the Site 658 sediments (Stabell, 1989). Mean accumulation rates of terrigenous organic matter reach low values of about 0.05 gC cm^{-2} ky^{-1}. This interval of major decrease in accumulation rates is interpreted as a transition from dominantly humid to more arid climatic conditions in the Central Sahara. The expansion of more arid conditions is also supported by an almost coeval increase in eolian dust supply recorded at DSDP-Site 397 further north (Fig. 57; Stein, 1985a).

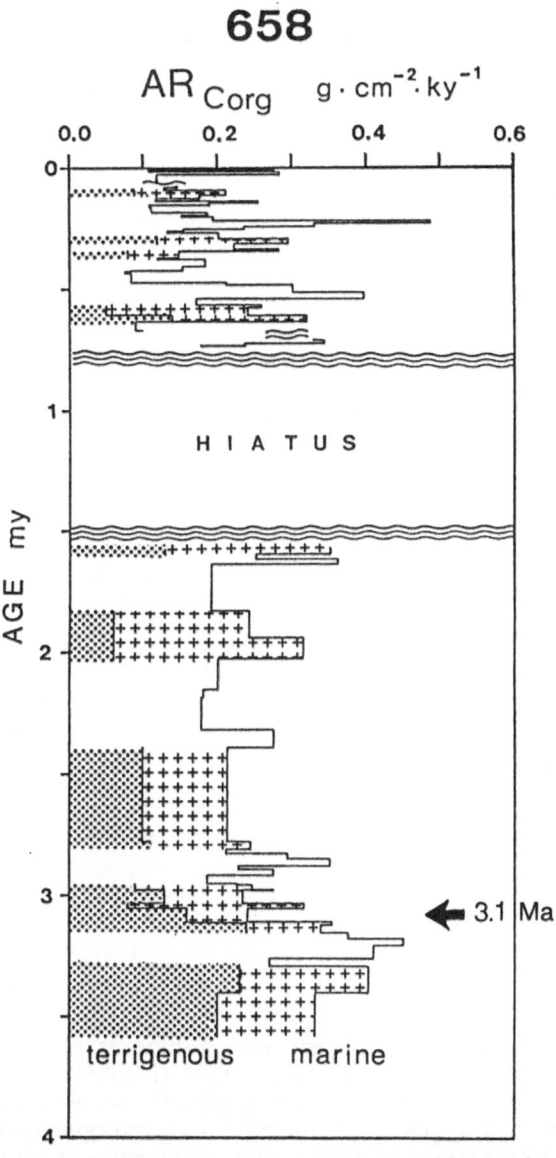

Fig. 83: Mean accumulation rates of terrigenous and marine organic carbon at Site 658, based on kerogen microscopy data (from Stein and Littke, 1990).

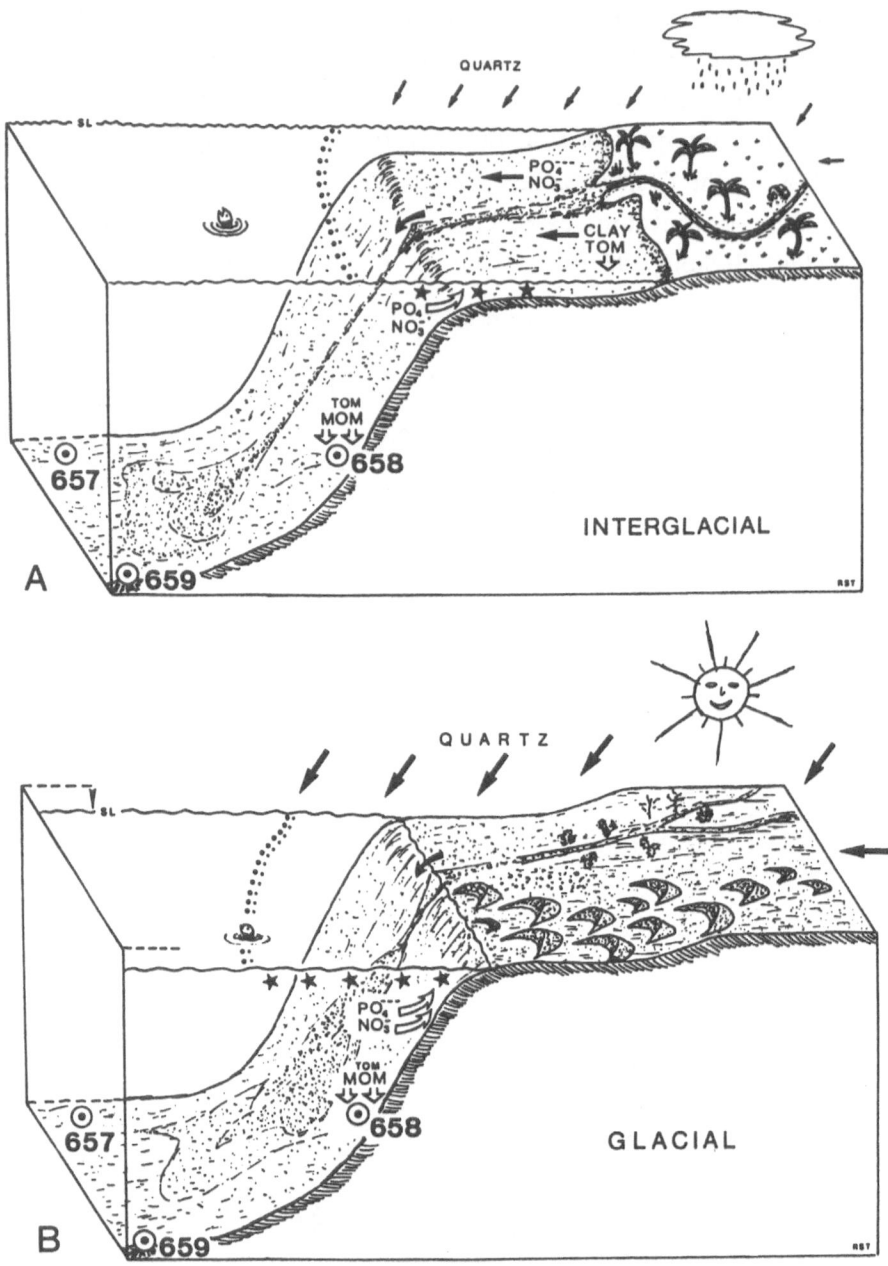

Fig. 84: Depositional model for Plio-Pleistocene sediments at the northwest African continental margin. A. Interglacial; B. Glacial. TOM (MOM) = terrigenous (marine) organic matter; ⤴= NE Trades (increased during glacials, reduced during interglacials); ⤴ upwelling (intensified during glacials, reduced during interglacials); ◀= fluvial input of clay fraction and nutrients (reduced during glacials because of arid climate, increased during interglacials because of humid climate); ★★ = area of high oceanic productivity. (For relationships between climate in Northwest Africa, trade wind intensity, and upwelling intensity, see Sarnthein et al., 1981, 1982; Stein, 1985a).

In the last 0.7 Ma, low accumulation rates of both terrigenous organic carbon and clay fraction generally coincided with glacial intervals, whereas the rates were distinctly increased during interglacials (Fig. 85). These high-amplitude short-term fluctuations are interpreted as a result of distinct cyclic changes in paleoclimate between arid and humid conditions in Northwest Africa.

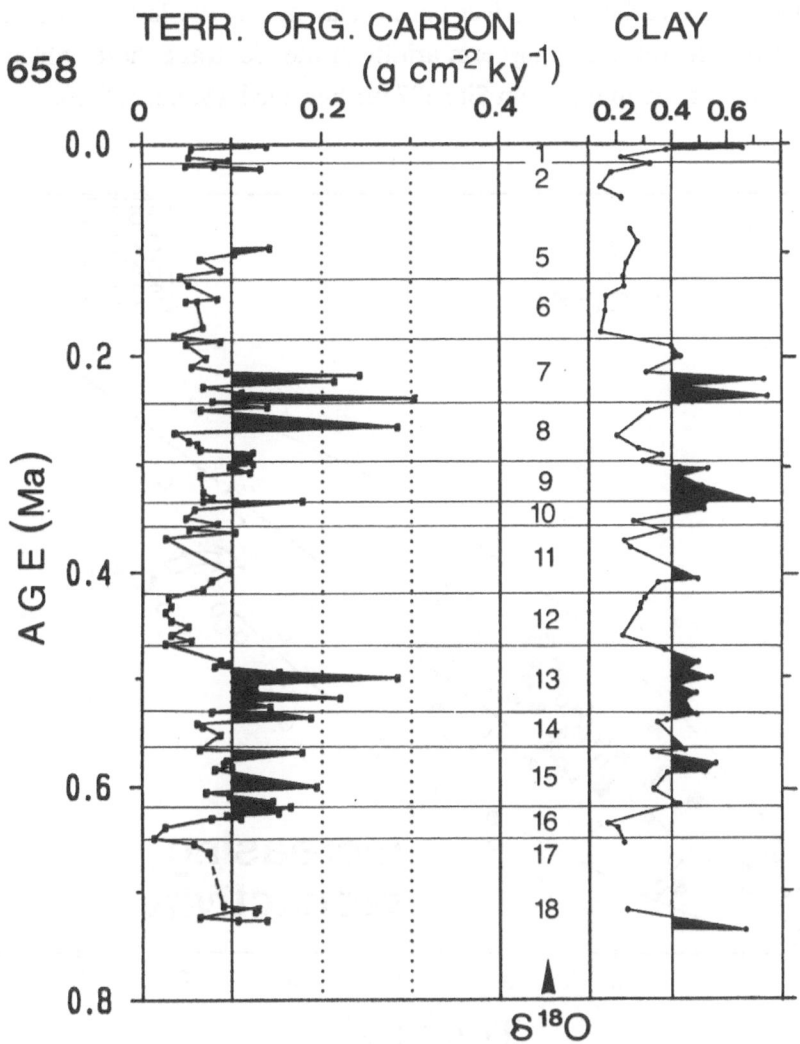

Fig. 85: Accumulation rates of terrigenous organic carbon and clay fraction for the last 0.75 Ma. Estimates of terrigenous organic carbon fraction based on hydrogen index values. Clay data and stable isotope stages from Sarnthein and Tiedemann, 1989 and Tiedemann et al., 1989).

During glacial times, an arid climate was dominant which caused a reduced fluvial contribution (i.e., decreased accumulation rates of clay and terrigenous organic matter) and an increased silt-sized eolian quartz input (Fig. 84B; Tiedemann et al., 1989). On the other hand, a more humid climate prevailed during interglacial times, resulting in low accumulation rates of (eolian) quartz and higher rates of supply of (fluvial) clays and terrigenous organic matter (Figs. 84 and 85; cf., Sarnthein and Koopmann, 1980; Stein, 1985a; Tiedemann, et al., 1989). The dominance of very low accumulation rates of both terrigenous organic carbon and (river-borne) clay during the last 0.5 Ma (Fig. 85) may suggest a further increase in long-term aridity in the Central Sahara, which is also indicated in the data for sediments from Site 397 further north (Stein, 1985a).

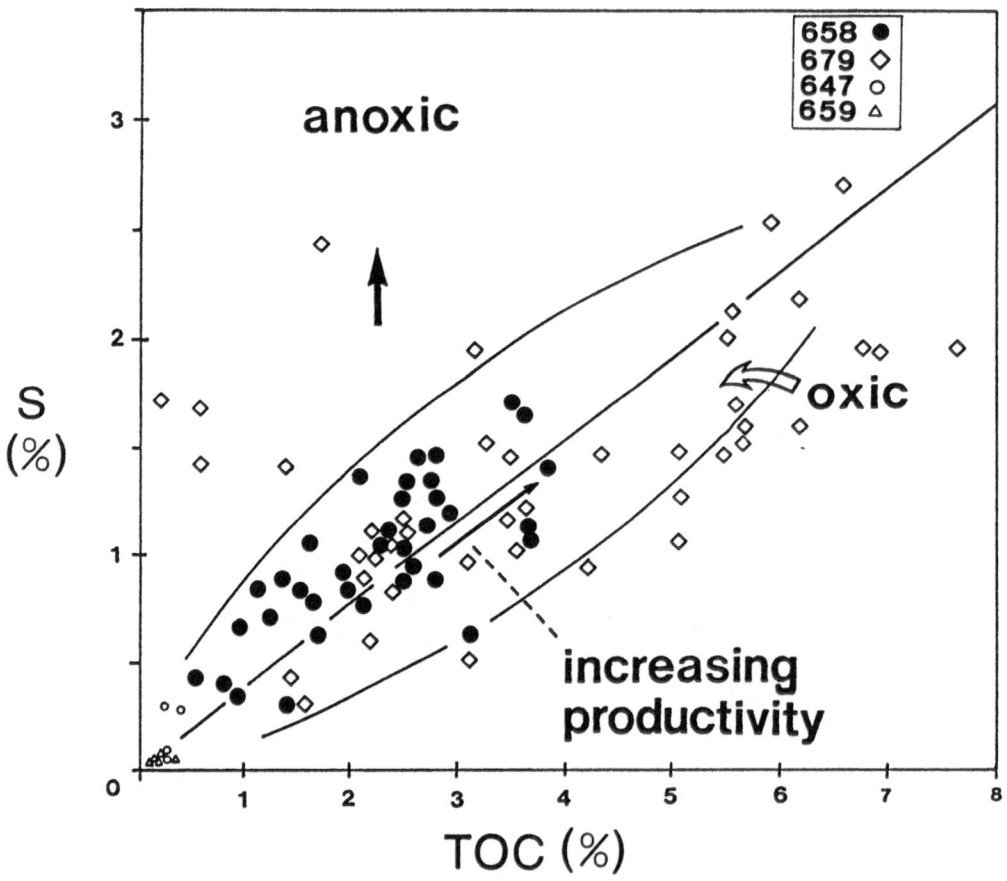

Fig. 86: Correlation between organic carbon and (pyritic) sulfur for sediments from Upwelling-Sites 658 (NW-Africa) and 679 (Peru) and open-ocean non-upwelling Sites 647 and 659 (cf., Fig. 16).

6.5.2. Marine organic carbon and paleoproductivity off Northwest Africa (Site 658)

Changes in accumulation rates of marine organic carbon may have resulted from changes in surface-water productivity and/or changes in preservation rate of organic matter (see Chapter 2.2). Since bioturbation was observed in most of the sequence of Site 658 (Ruddiman, Sarnthein, et al., 1988), an anoxic deep-water environment with increased preservation of organic matter is unlikely. The correlation between organic carbon content and sulfur also argues against anoxic deep-water conditions (Fig. 86; cf., Leventhal, 1983). Thus, the changes in marine organic carbon deposition are interpreted in terms of changes in surface-water productivity.

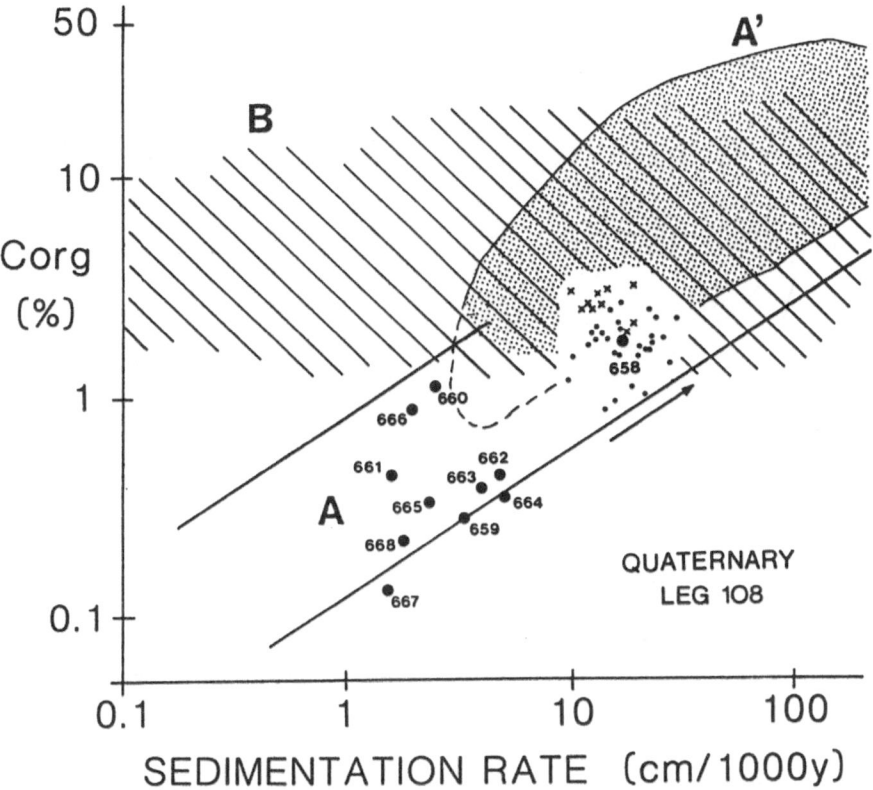

Fig. 87: Correlation between (marine) organic carbon content and sedimentation rate for Leg 108 sediments (according to Stein, 1986a, 1990). For further explanation of fields A, A', and B see Figure 11.

In an oxic marine environment, average surface-water paleoproductivity can be roughly estimated from the relationship between (marine) organic carbon content and sedimentation rate (Müller and Suess, 1979; Stein, 1986b, 1990). The sediments at all sites occupied during ODP-Leg 108, except Site 658, are characterized by relatively low organic carbon contents and low sedimentation rates (Fig. 87), indicating an oxic low-productivity environment typical of the open ocean. At Site 658, on the other hand, high organic carbon contents and high sedimentation rates (Fig. 87) point to a high-productivity environment typical of coastal upwelling areas. The situation is similar for the upwelling areas off Peru and Arabia (Fig. 88; Stein, 1990).

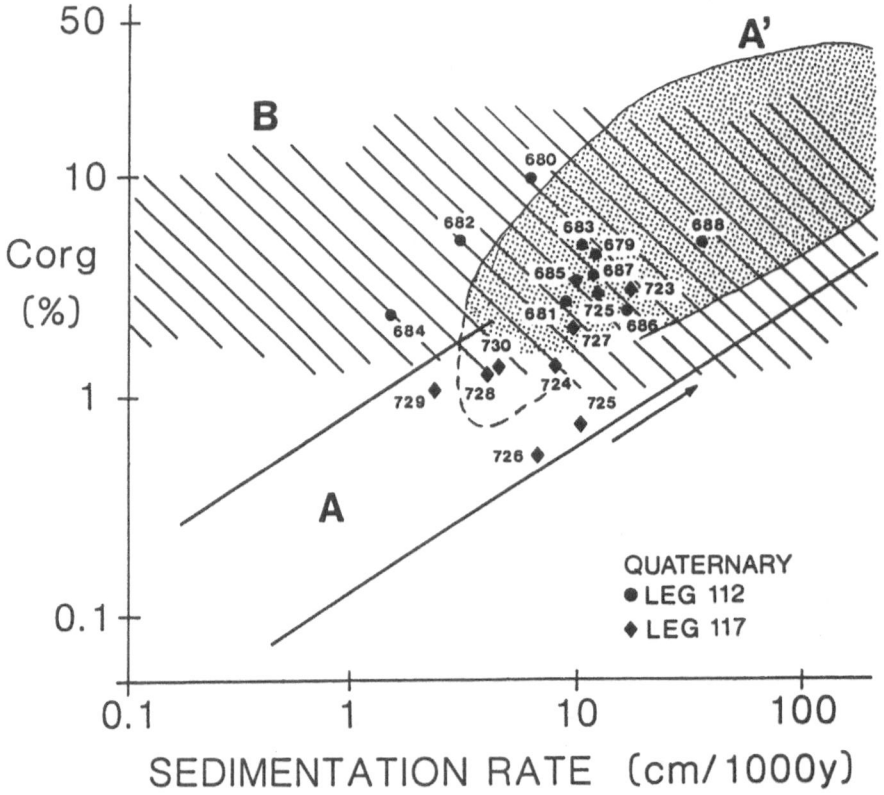

Fig. 88: Correlation between (marine) organic carbon content and sedimentation rate for Leg 112 and 117 sediments (according to Stein, 1986b, 1990; Data from Prell, Niitsumo et al., 1988; Suess, von Huene et al., 1988). For further explanation of fields A, A', and B see Figure 11.

For a more detailed reconstruction of the history of surface-water productivity off Northwest Africa, however, accumulation rates of (marine) organic carbon and biogenic opal, paleoproductivity records (cf. Chapter 3.1.), as well as a detailed oxygen isotope stratigraphy have to be used (cf., Ruddiman et al., 1989; Stein et al., 1989b; Tiedemann et al., 1989). It is of particular interest to know (1) whether high productivity off Northwest Africa at the location of Site 658 was caused by increased coastal upwelling and/or by increased fluvial nutrient supply (cf., Diester-Haass, 1983; Stein, 1985a) and (2) when increased productivity occurred (i.e., during glacial or interglacial intervals).

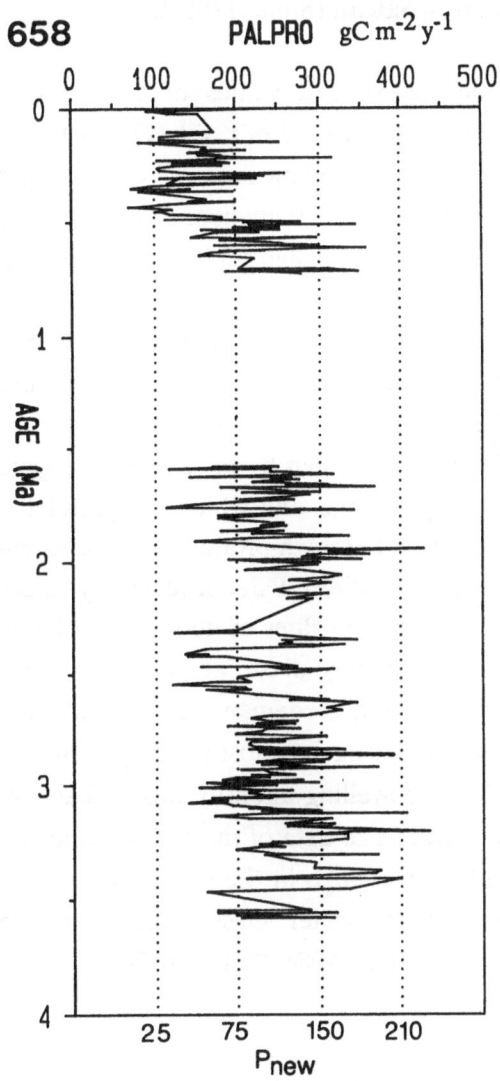

Fig. 89: Paleoproductivity estimates for Site 658 sediments, calculated using equation (4).

The distinct increase in biogenic opal accumulation rates near 3 Ma is interpreted as the onset of the long-term persistent upwelling activity off Cap Blanc (Mienert et al., 1988; Tiedemann et al., 1989). This increased upwelling of cold waters driven by strongly enhanced trade winds may have caused the increase in aridity in the central Sahara. This is indicated by the almost contemporaneous decrease in accumulation rates of (river-borne) clay and terrigenous organic matter (Stein et al., 1989b; Tiedemann et al., 1989). Both increased eolian sediment supply and biogenic opal content, also recorded at DSDP-Site 397 further north, corroborate the interpretation of increased upwelling and increased aridity in the central Sahara (Stein, 1985a).

Estimates of paleoproductivity yielded values of 180 to 400 gC m^{-2} y^{-1} or new production values of about 60 to 210 gC m^{-2} y^{-1} (Fig. 94) which are similar or even higher to those recorded in the upwelling area off Cap Blanc today (Fig. 5, Tab. 1). There is no long-term change in productivity near 3 Ma. That means, that an increased fluvial nutrient supply may have caused high surface-water productivity prior to 3 Ma.

During the late Pliocene/Pleistocene time interval, most maxima in biogenic opal accumulation occur contemporaneous with maxima in marine organic carbon accumulation rates (Ruddiman et al., 1989). Distinct peaks in biogenic opal accumulation of 7 to 17 g m^{-2} y^{-1} which by far exceed the generally high opal values known from coastal upwelling regimes and which are paralleled by maxima in (river-borne) clay input, are interpreted as high-productivity events caused by special fluvial nutrient supply (Tiedemann et al., 1989). Paleoproductivity values may exceed values of 300 gC m^{-2} y^{-1} (i.e., new production values of about 150 gC m^{-2} y^{-1}) (Fig. 89). During the last 0.7 Ma, these events mainly occur at peak and early warm stages and at terminations (Fig. 90; Sarnthein and Tiedemann, 1989; Tiedemann et al., 1989). On the other hand, productivity was lower during most of the glacial intervals such as stage 10, 12, 14, and 16 (Fig. 90). Since upwelling intensity was probably increased during glacial intervals as suggested from grain-size data of the eolo-marine sediment fractions at Site 658 (Tiedemann et al., 1989), a change in the source water of coastal upwelling from nutrient-rich to less nutrient-rich water masses may also explain the decrease in productivity (cf., Stein and Sarnthein, 1984; Stein, 1985a).

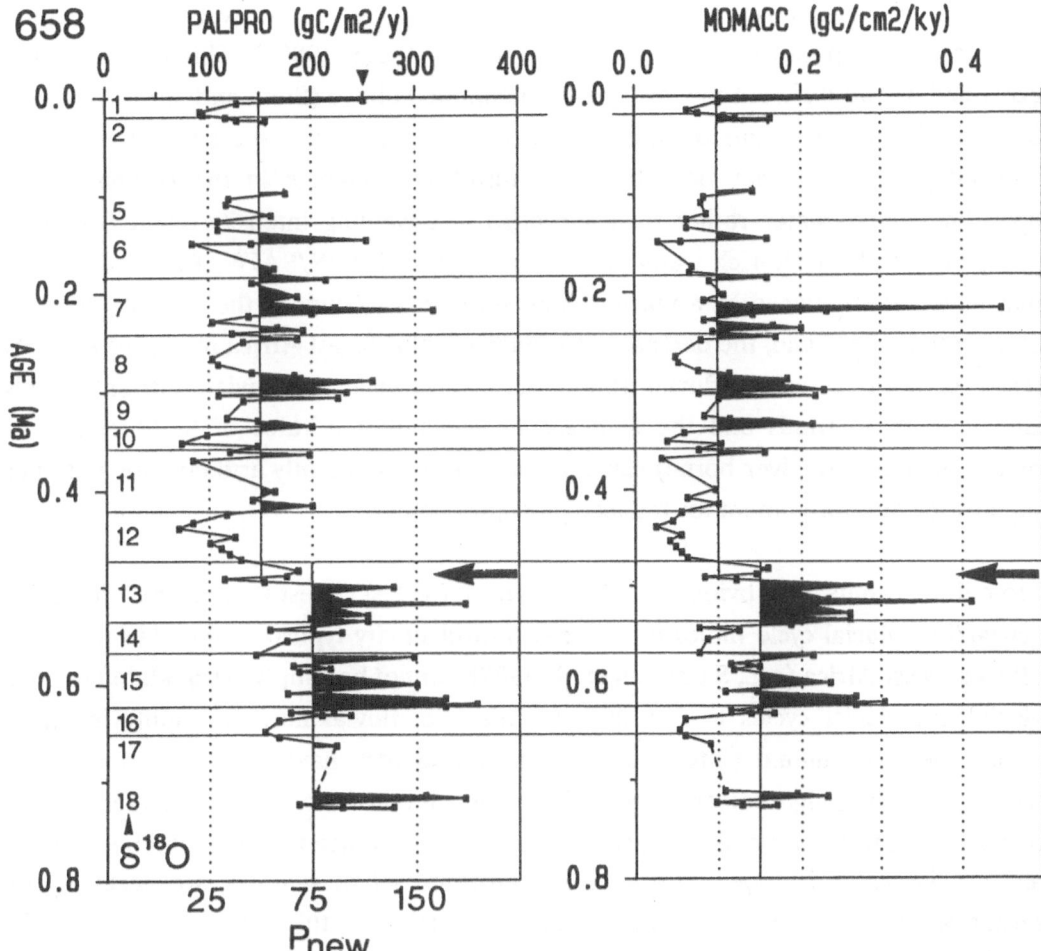

Fig. 90: Estimates of paleoproductivity calculated using equation (4), and accumulation rates of marine organic carbon for the last 0.75 Ma. Oxygen isotope stages according to Sarnthein and Tiedemann (1989). Solid triangle indicates recent surface-water productivity measured off Cap Blanc (Schemainda et al., 1975). Solid arrow indicates long-term change in productivity near 0.5 Ma.

Near 0.5 Ma, long-term paleoproductivity distinctly decreased (Fig. 90). A similar decrease contemporaneously occurred at Site 397 (Stein, 1985b). This change can be explained by two different models already mentioned above (Stein and Sarnthein, 1984; Stein, 1985a). Since grain-size data suggest enhanced upwelling throughout the interval, reduced productivity may have been the result of a change in the composition of upwelling source water rather than a change in upwelling intensity. Near 0.5 Ma, a phosphate- and nitrate-rich water mass (such as the SACW/SAIW; Bainbridge 1976) may have been replaced by a nutrient-poor water mass (such as the NACW) (cf., Fig. 60). On the other hand, the decrease of fertility off Northwest Africa may have been the result of a drying-up of fluvial runoff, i.e., a stoppage of fluvial nutrient supply. Enhanced trade winds and the decrease in accumulation rates of both terrigenous organic matter and (river-borne) clay may indicate a dominantly arid climate and, thus, support the latter argument for reduced paleoproductivity.

From these results it is obvious that the data from the last (latest Pleistocene/Holocene) glacial/interglacial cycle indicating increased productivity typical of the glacial interval off Northwest Africa (e.g., Sarnthein et al., 1987), cannot be simply extrapolated to older glacial/interglacial cycles at Site 658. At this Site, fluvial nutrient supply increased during the more humid (interglacial) intervals and may have become the dominant factor controlling high paleoproductivity. Further detailed work on accumulation rates of marine organic carbon and biogenic opal, foraminifera assemblages, and both carbon and oxygen stable isotopes performed on the same set of samples, are necessary for reconstruction of the history of surface-water productivity in the Northeast Atlantic. This future work is also important for global reconstructions of changes in productivity and CO_2 budget and for climate modelling.

6.5.3. Turbidites and organic carbon preservation in the continental rise off Northwest Africa (Sites 657 and 659)

At Site 657, normal pelagic sedimentation is often interrupted by events of turbidity currents and slumping (Fig. 91; Faugères et al., 1989a). Similar events, although less frequent, also occurred at Site 659 (Fig. 92; Faugères et al., 1989b). Besides their typical characteristics in color, grain size, and texture (Faugères et al., 1989a, b), most of the redeposited sequences can be distinguished from the normal pelagic sediments by their high organic carbon contents (1 - 3%) and relatively low carbonate contents (20 - 45%) (Figs. 66, 92, and 93; Tab. 7). Such phenomena were also described for adjacent deep-sea basins in the eastern subtropical Atlantic (Weaver and Kuijpers, 1983; Colley et al., 1984).

The quality of the organic matter in the slump interval and in the fine-grained thin-bedded turbidites is very similar to that found in the organic-carbon-rich sediments at Site 658, as indicated by Rock-Eval pyrolysis, $\delta^{13}C_{org}$, and C/N data (Figs. 75 and 93, Tab. 6; Stein et al., 1989b). The sediments of the slump are furthermore characterized by high amounts of marine diatoms (Stabell, 1989). Thus, the source area of both these turbidites and the slump is very likely to be the upwelling-influenced, high-productivity area of the upper slope (Fig. 84). Since the organic matter in turbidites and slumps was rapidly buried after redeposition in the deep sea, the residence time in zones of bioturbation and oxic decomposition was short. Consequently, relatively high amounts of marine (hydrogen-rich) organic matter could be preserved in an oxygen-rich, open-marine environment.

However, not all of the turbidites have high organic-carbon contents. A few of the turbidites at Site 657 are coarse-grained and thick-bedded with a thickness of up to more than 3 meters (Faugères et al., 1989a). One of these turbidites is documented among the sediment samples investigated for organic carbon content (Tab. 7). The organic carbon values are very low (less than 0.1 %) whereas carbonate contents are very high (up to 85 %). Furthermore, the amount of coarse fraction which is mainly composed of quartz, foraminifera, and shell fragments (cf., also Faugères et al., 1989a), may reach values of up to 80 % (Tab. 7). According to these data, the source of the coarse-grained turbidites is very different from that of the fine-grained, organic-carbon-rich turbidites decribed above; it is probably the shelf area off Northwest Africa.

130

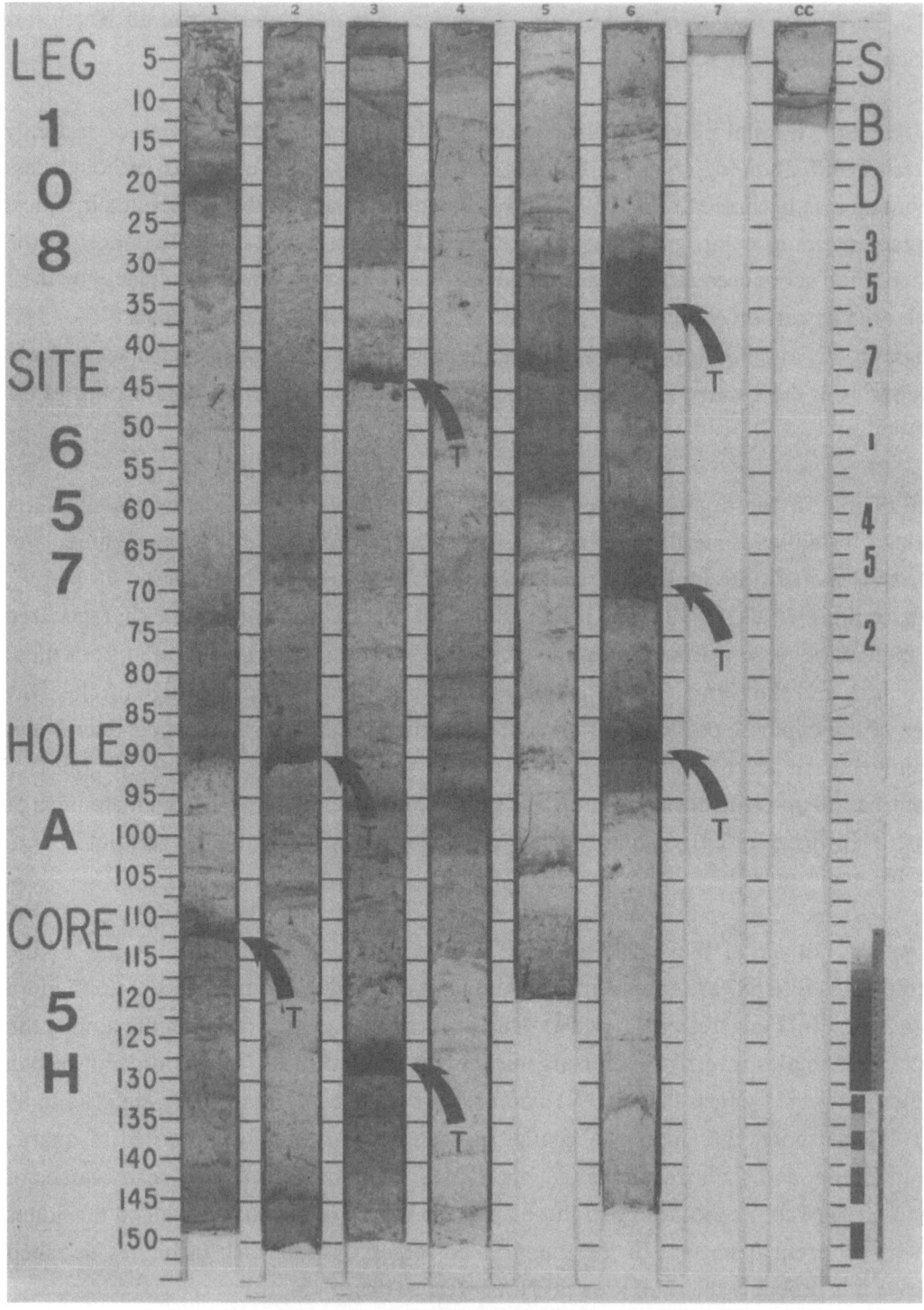

Fig. 91: Photograph of Core 657A-5H, showing several turbidites (T).

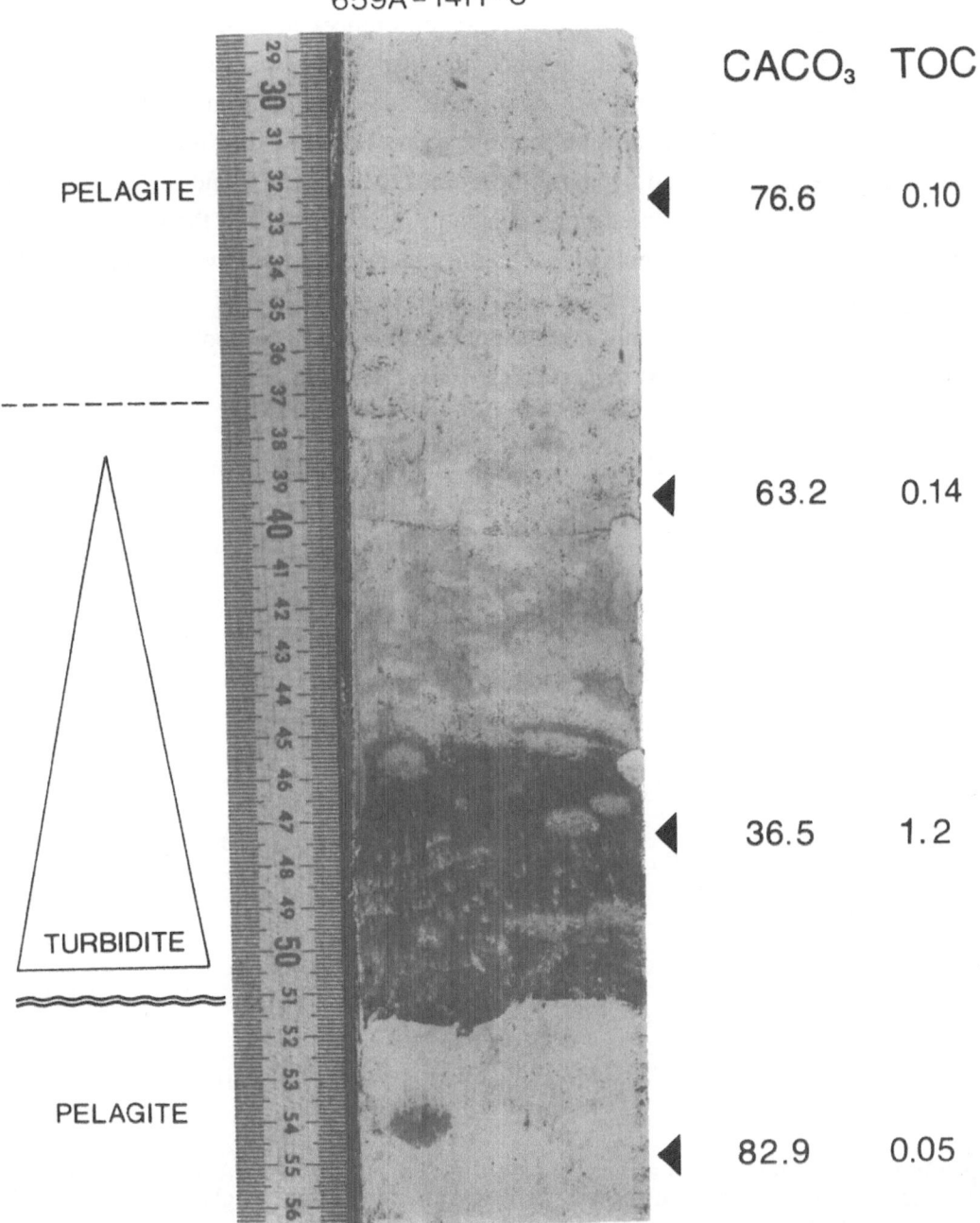

Fig. 92: Photograph and total organic carbon and carbonate data (%) of a turbidite in Section 659A-14H-3.

According to the detailed sedimentological study of the slumped interval between 12 and 27 mbsf (cf., Fig. 66) by Günther (1988), differences in composition and, thus, source area of the redeposited sediments are also obvious. This interval can be separated into two turbidites and two slumps which are very different in their organic-carbon contents and their amounts of coarse fraction (Fig. 94). The turbidites and the upper slump are characterized by high TOC values of 1 to 2.5 % and low amounts of coarse fraction (about 10 %) indicating the high-productivity zone of the upper slope to be the source area of these sediments. On the other hand, the lower slump displays distinctly lower TOC values of less than 0.5 % and high amounts of coarse fraction of about 40 %. The coarse fraction of these sediments are mainly composed of quartz and shell fragments (Günther, 1988) suggesting an origin from the shelf.

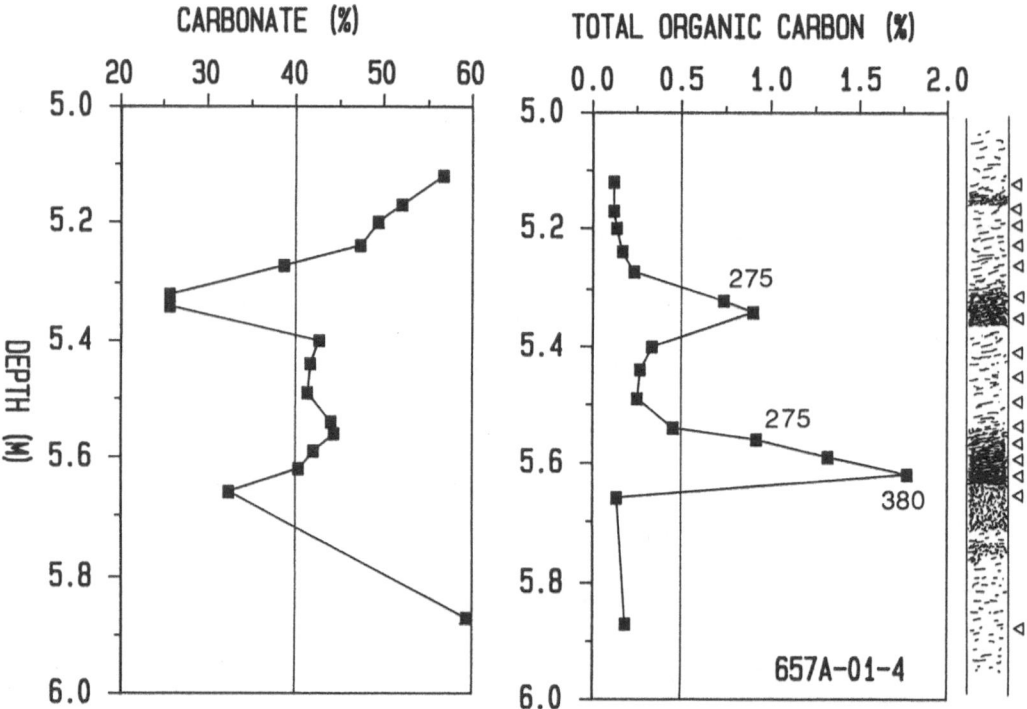

Fig. 93: Total organic carbon and carbonate data (%) of Core 657A-1H-4, 60-140 cm. Numbers in the organic carbon record are hydrogen index values (mgHC/gC) measured on carbonate-free samples.

Nevertheless, the dominant portion of the redeposited sediments at Site 657 is enriched in organic carbon and accounts for about 15 % of the upper 100 m of the total sediment sequence. That means, the storage of organic carbon in the deep-ocean basins by this process may also be important for calculation of global carbon budgets as well as for economic aspects (i.e., the formation and prospection of petroleum source rocks).

Tab. 7: Carbonate and organic carbon contents, and amounts of coarse fraction in turbidite sequences (T) at Site 657.

Sample	Depth (m)	CaCO$_3$ (%)	C$_{org}$ (%)	> 63 μm (%)	
01-4-062	5.12	56.7	0.11	n.d.	
01-4-067	5.17	52.0	0.12	n.d.	
01-4-070	5.20	49.4	0.14	14.6	
01-4-074	5.24	47.5	0.16	5.6	
01-4-077	5.27	38.6	0.23	13.3	
01-4-082	5.32	25.6	0.74	0.4	} T
01-4-084	5.34	25.6	0.90	n.d.	
01-4-090	5.40	42.6	0.34	4.5	
01-4-094	5.44	41.7	0.26	4.3	
01-4-099	5.49	41.2	0.25	3.7	
01-4-104	5.54	44.0	0.45	0.2	
01-4-106	5.56	44.3	0.92	n.d.	} T
01-4-109	5.59	41.9	1.32	n.d.	
01-4-112	5.62	40.3	1.76	0.2	
01-4-116	5.66	32.3	0.13	5.6	
01-4-137	5.87	59.4	0.18	n.d.	
05-3-120	39.90	47.9	0.14	n.d.	
05-3-121	39.91	53.5	0.29	5.6	
05-3-126	39.96	28.6	1.44	0.2	
05-3-128	39.98	24.7	0.66	n.d.	} T
05-3-130	40.00	27.3	0.44	0.4	
05-3-133	40.03	21.9	0.55	0.4	
05-3-136	40.06	20.5	0.26	0.8	
09-3-002	76.72	84.6	0.07	32.5	
09-3-010	76.80	76.0	0.03	80.9	
09-3-015	76.85	80.3	0.05	77.5	
09-3-017	76.87	78.1	0.00	n.d.	} T
09-3-020	76.90	75.9	0.04	78.4	
09-3-025	76.95	80.9	0.05	80.9	
09-3-030	77.00	75.3	0.06	79.5	
09-3-035	77.05	71.7	0.08	57.8	

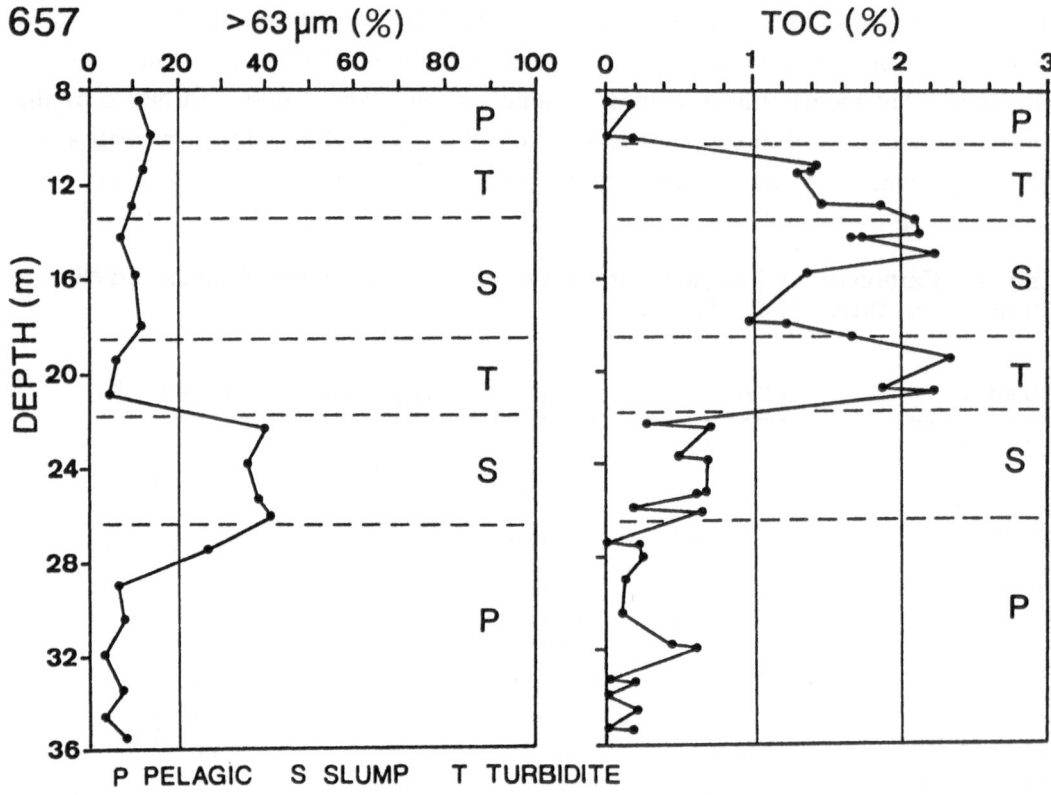

Fig. 94: Total organic carbon content and coarse fraction of sediments of the slumped interval (about 10 to 27 mbsf) of Site 658 (after Günther, 1988).

6.5.4. Organic carbon and carbonate cycles at Site 660: Productivity and/or deep-water circulation signal ?

At Site 660, high-amplitude variations in organic carbon and carbonate contents are well preserved in the upper Pliocene and Quaternary sediments (Fig. 68). These fluctuations can either be the result of changes in supply of carbonate and organic carbon or the result of dilution by other sediment components such as siliciclastics. Since Site 660 lies on the Sierra Leone Rise, far away from the continent, major changes in the terrigenous supply causing fluctuations in carbonate content because of dilution are very unlikely (cf., Dean et al., 1981; Stein, 1985b). Thus, no negative correlation exists between accumulation rates of non-carbonates (i.e., terrigenous matter) and the carbonate content (Fig. 95). This is very different from the situation in the hemipelagic sediments at Site 658, where distinct dilution by terrigenous matter caused reduced carbonate contents (Fig. 65 and 95). In parts of the late Pliocene, however, dissolution of carbonate may have become important also at Site 658 which is corroborated by decreased preservation of foraminifera (Weaver and Raymo, 1989).

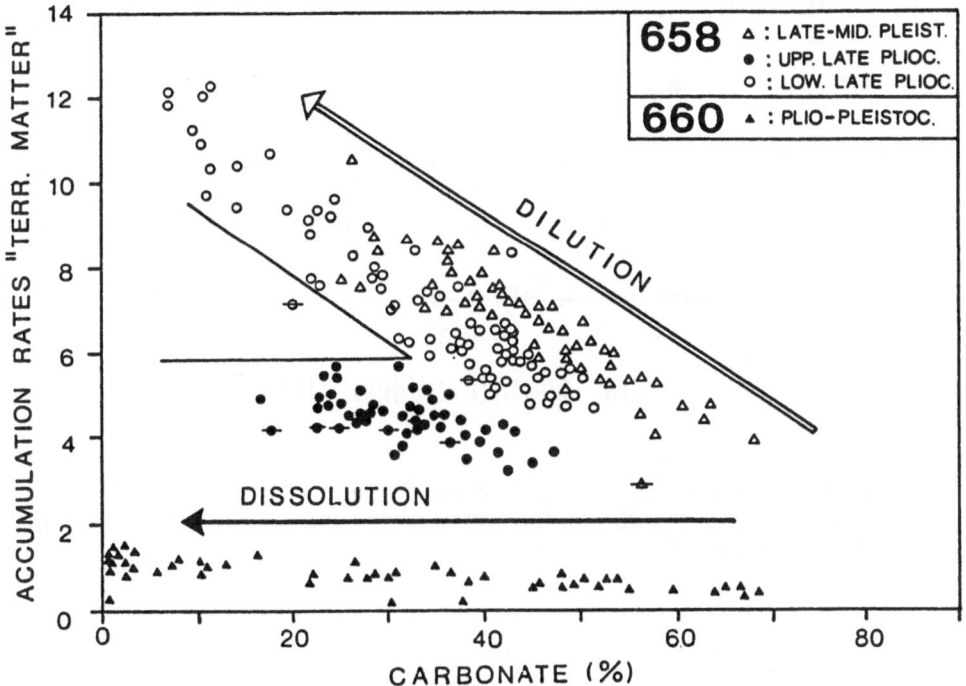

Fig. 95: Accumulation rates of terrigenous (siliciclastic) matter (g cm^{-2} ky^{-1}) (based on mean sedimentation rates) vs. carbonate content at Sites 658 and 660.

At Site 660, on the other hand, high carbonate values coincide with low organic carbon contents and vice versa (Fig. 96). This may suggest a relative enrichment of organic carbon because of dissolution of carbonate. Strong carbonate dissolution is also evident from the degree of preservation of foraminifera at Site 660 (Weaver and Raymo, 1989). Because dilution effects by siliciclastics can be excluded, the carbonate/organic carbon relationship of Figure 96 is mainly controlled by changes in carbonate and/or organic-carbon fluxes (cf., Ricken, 1989).

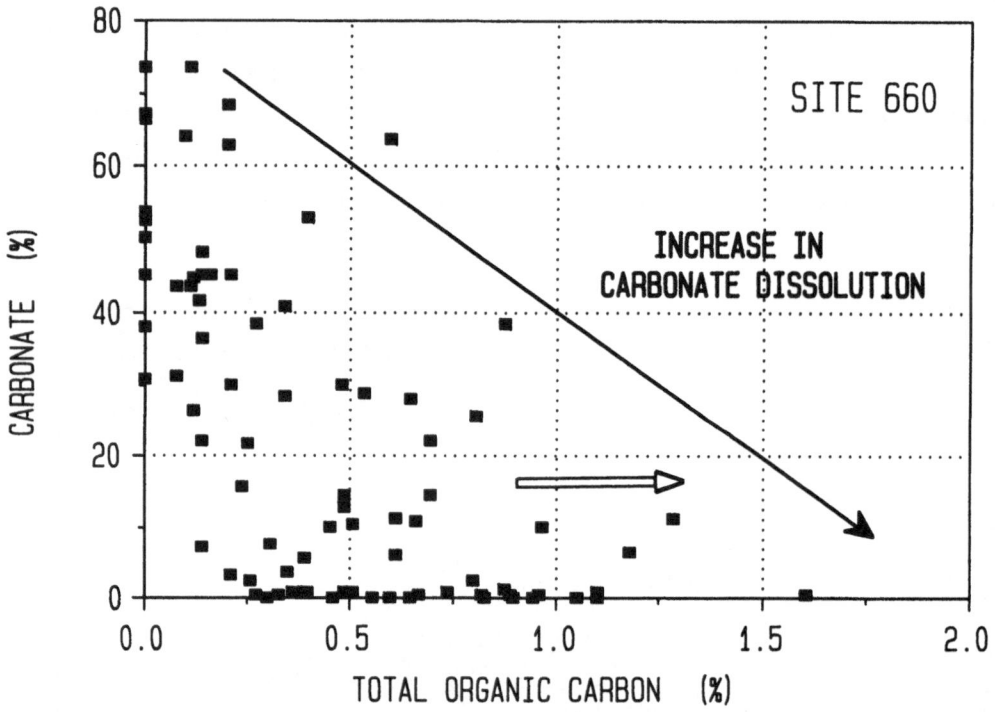

Fig. 96: Carbonate content vs. total organic carbon content at Site 660. Solid arrow indicates increase in carbonate dissolution; open arrow indicates increase in organic carbon supply/preservation.

Carbonate variations between 70 and 0 % as recorded at Site 660, and probably caused by carbonate dissolution, can explain the late/middle Pleistocene organic carbon fluctuations between 0.5 and 1.7 %, without any change in the flux (production) rate of organic matter (Fig. 97 ; cf., Ricken, 1989). However, major dissolution of carbonates would also result in a distinct decrease in sedimentation rate (Fig. 97). Taking the example of Figure 97 and the organic carbon/sedimentation rate relationship (see Chapter 2; Stein, 1990), this would require an oxygen-deficient environment with increased preservation rate of organic matter to explain the maximum TOC values at Site 660 (Fig. 98). Microlaminations preserved because of the absence of bioturbation and common in the dark carbonate-lean intervals (Ruddiman, Sarnthein et al., 1988), point to an oxygen-deficient bottom-water environment (Fig. 99). Increased preservation of marine hydrogen-rich organic matter under these conditions occurring during times of maximum carbonate dissolution, may explain the relationship between carbonate content and hydrogen index (Fig. 100; cf., Demaison and Moore, 1980; Pratt, 1984; Canfield, 1989).

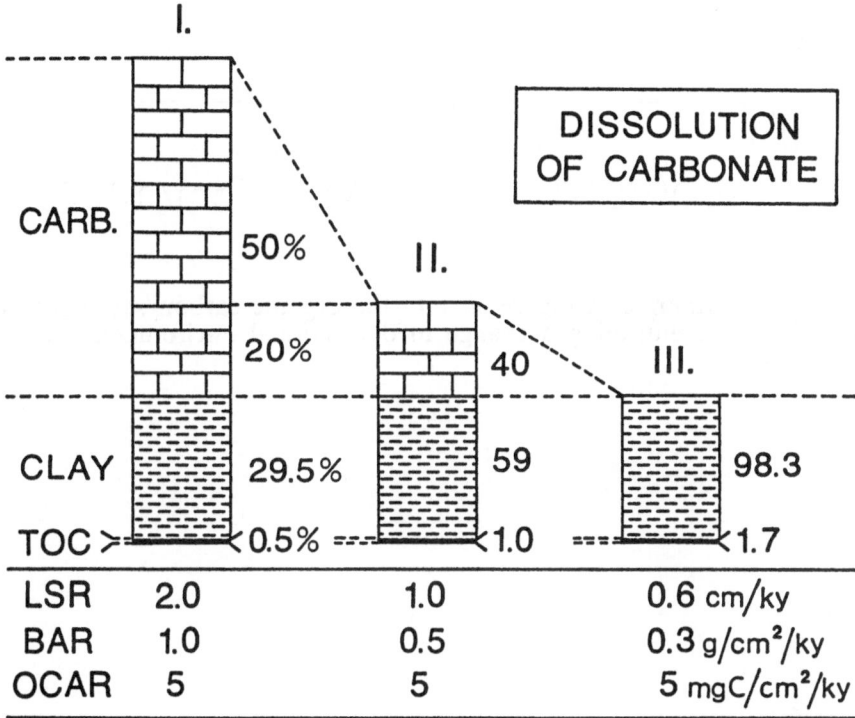

Fig. 97: Example for changes in TOC percentages (assuming constant organic carbon accumulation rates) caused by increased carbonate dissolution. LSR = linear sedimentation rate; BAR = bulk accumulation rate; OCAR = organic carbon accumulation rate.

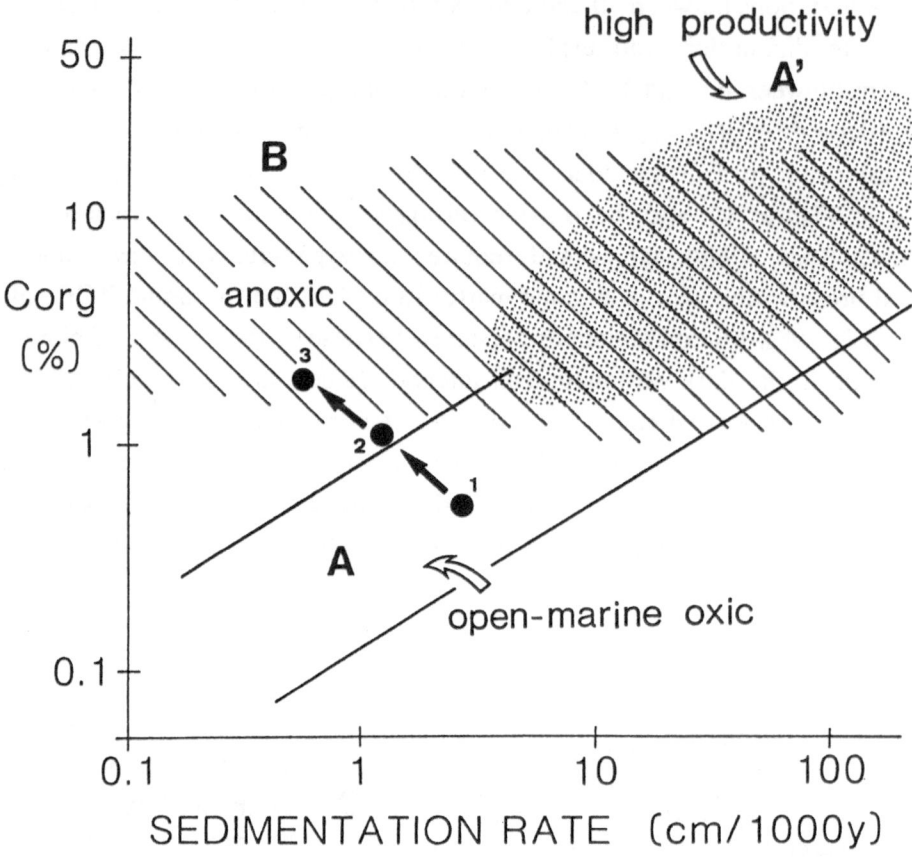

Fig. 98: Plot of the example of Figure 97 into the organic carbon/sedimentation rate diagram (cf., Fig. 11), indicating a change in depositional environment from oxic to anoxic conditions.

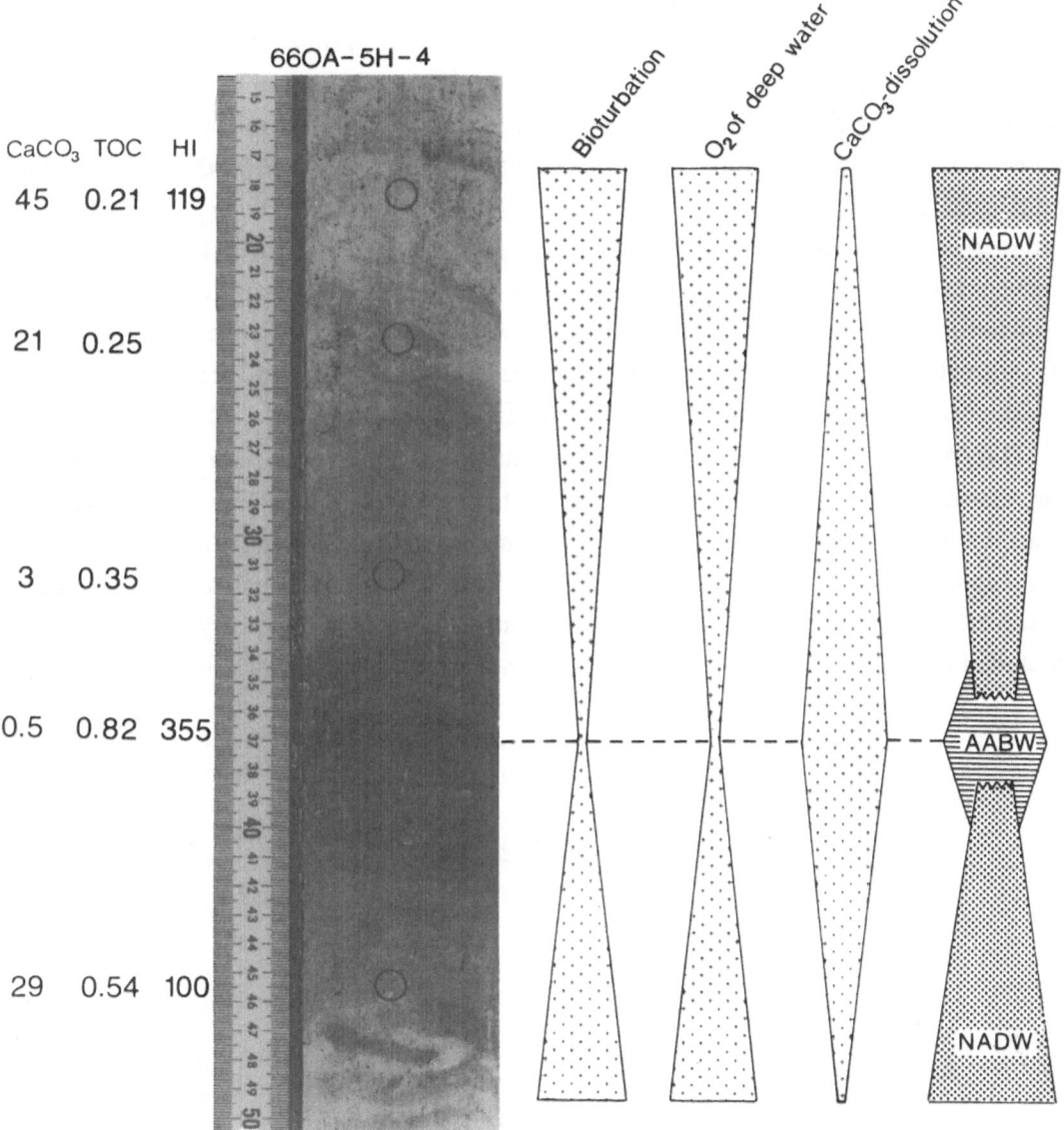

Fig. 99: Changes in total organic carbon and carbonate contents (%), hydrogen indices (mgHC/gC), and degree of bioturbation in Section 660A-5H-4, and the interpretation in terms of oxygenation of deep water, carbonate dissolution, and influence of NADW/AABW.

This interpretation would support the model of Curry and Lohmann (1983, 1985) who postulated an increased carbonate dissolution during glacial times because of a reduced production rate of well oxygenated NADW and increased influence of the oxygen-poor and carbonate-aggressive AABW in the late Quaternary eastern tropical Atlantic. Based on oxygen and carbon isotope variations determined in benthic foraminifera from the equatorial Atlantic Site 665 (Fig. 57), Curry and Miller (1989) suggested that the reduced oxygen content of the deep water was probably associated with low $\delta^{13}C$ values and contributed to increased preservation of organic carbon during cold (i.e., $\delta^{18}O$ enriched) intervals of the Pliocene equatorial Atlantic. This means that during (glacial ?) intervals characterized by carbonate contents around zero and increased (marine) organic carbon contents, the sediment/water interface at Site 660 lay below the NADW/AABW boundary and thus in the influence of the oxygen-poor and carbonate-aggressive AABW whereas during times of high carbonate and low organic carbon preservation this interface was above this boundary and in the influence of the well-oxygenated NADW (Fig. 99 and 100; cf. also Emerson, 1985). It has to be mentioned, however, that glacial/interglacial changes in surface-water productivity should not be excluded as further mechanism causing changes in organic carbon variations.

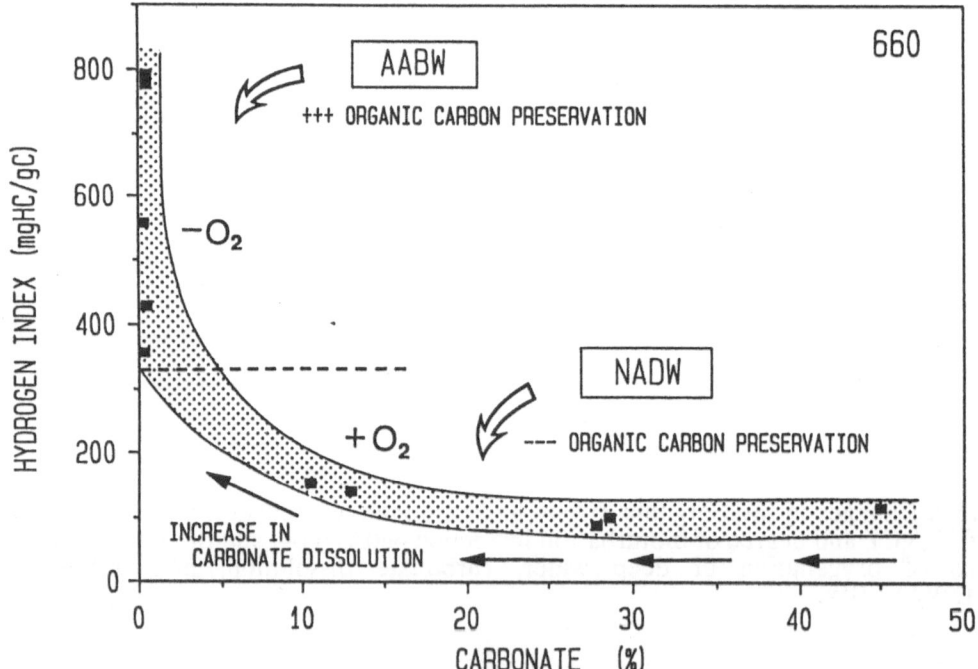

Fig. 100: Carbonate content vs. hydrogen index at Site 660 and the interpretation in terms of changes in oxygen content of deep water, preservation of organic carbon, dissolution of carbonate, and influence of NADW/AABW.

Nevertheless, an increased surface-water productivity because of increased upwelling in the North-Equatorial Divergence Zone (Fig. 58) cannot be the only explanation for increased TOC values at Site 660, because at the close-by Site 661 no enrichment of organic carbon was recorded. The model described above, however, can also explain these differences: In contrast to the deeper Site 660, the shallower sediment/water interface at Site 661 probably lay above the NADW/AABW boundary throughout the time intervals represented in the investigated sediments, so that the well-oxygenated NADW prevented preservation of organic carbon (Fig. 60).

On the other hand, the distinct increase in organic carbon accumulation rates near 2.4 Ma with maximum values in the late/middle Pleistocene at Site 660, may point to an increase in productivity. This increase in accumulation rates of organic carbon in the North-Equatorial Divergence Zone near 2.4 Ma parallels a similar increase recorded at the Equatorial-upwelling Site 662 (Figs. 69 and 71). At the latter site, however, the increase in accumulation rates of organic carbon coincides with increased accumulation rates of biogenic opal, which is interpreted as a signal for higher surface-water productivity caused by increased equatorial upwelling because of intensified meridional atmospheric circulation (Fig. 69; Ruddiman and Janecek, 1989). A similar phenomenon (i.e., a decrease in carbonate and an increase in organic carbon and biogenic opal contents, and in sedimentation rate) was also observed in a piston core from the Guinea Basin ($4^{\circ}56.8'S$, $0^{\circ}27.7'E$; 4735 m water depth) and 'interpreted to reflect increased productivity in the equatorial upwelling belt (Morris et al., 1984). At Site 660, no such increase in biogenic opal was observed in the Plio-Pleistocene time interval (Ruddiman, Sarnthein et al., 1988). Furthermore, when interpreting the accumulation rate record of Site 660, it has to be considered that average sedimentation rates were used to calculate accumulation rates. For example, an average rate of 3.2 cm/ky was used for both carbonate-lean and carbonate-rich intervals from the upper 14 m of the sediment sequence at Site 660. According to the example in Figure 97, however, a constant sedimentation rate for these intervals is unlikely. On the other hand, distinct short-term changes in sedimentation rates are less probable for Site 662 because of the absence of significant carbonate dissolution (Weaver and Raymo, 1989).

From the discussion above, it is obvious that the interpretation of the organic carbon data from Site 660 is still preliminary. In order to interpret the organic carbon data from Site 660 (and Site 661) more precisely, i.e., to decide whether an increased preservation rate or an increased production rate of organic carbon was the dominant factor controlling organic carbon enrichment, more informations from both sites are required, such as:

(1) a detailed oxygen-isotope stratigraphy (like that available for Site 658; Sarnthein and Tiedemann, 1989) to calculate sedimentation rates for (carbonate-lean) glacial and (carbonate-rich) interglacial intervals separately;

(2) more detailed organic carbon records for both sites;

(3) more data about the composition of the organic matter deposited at Sites 660 and 661 during glacial/interglacial times;

(4) more detailed data on biogenic opal content.

6.6. Conclusions

The detailed sedimentological and organic-geochemical investigations of Leg-108-sediments yielded important insights into changes in paleoclimate and paleoceanic circulation and into the mechanisms controlling the accumulation of organic carbon in the tropical/subtropical eastern Atlantic Ocean. The data can be regarded as an example/model for organic-carbon enrichment by increased surface-water productivity in upwelling areas, increased preservation rate, and/or increased supply of terrigenous sediment fraction. The results can be summarized as follows:

(1) The sediments from coastal-upwelling Site 658 are characterized by high organic carbon contents of up to 4% and organic carbon accumulation rates of up to 0.4 $gC\ cm^{-2}\ ky^{-1}$. The organic matter is a mixture of marine and terrigenous material, with a dominance of the marine proportion. At Site 660 (Northern Equatorial Divergence Zone) and Site 662 (Equatorial Divergence Zone), up to 1.5 % of (marine) organic carbon was recorded. In contrast to this, the pelagic sediments from non-upwelling Sites 657 and 659 display low organic carbon contents of less than 0.5 % and very low accumulation rates of less than 0.01 $gC\ cm^{-2}\ ky^{-1}$. At the latter two sites, however, turbidites and slumps are intercalated and characterized by organic carbon contents of up to 3%.

(2) The distinct decrease in accumulation rates of terrigenous organic matter and (river-borne) clay at Site 658 between 3.1 and 2.8 Ma indicates the change from a dominantly humid climate in the Central Sahara prior to 3.1 Ma to a climate of arid/humid (semi-arid) cycles typical of the subsequent late Pliocene/Pleistocene interval. During glacial intervals, the flux rates of both terrigenous organic matter and (river-borne) clay were

low because of a dominantly arid climate, whereas during the more humid interglacials the flux rates were distinctly increased. Near 0.5 Ma, a further increase in long-term aridity in the Central Sahara probably occurred.

(3) Paleoproductivity was generally high in the upwelling area off Northwest Africa during the last 3.6 Ma reaching values between 100 and 400 gC m^{-2} y^{-1}. Most of the higher productivity values were recorded at peak interglacials and at terminations, indicating the importance of fluvial nutrient supply at Site 658. Near 0.5 Ma, long-term productivity distinctly decreased which can be explained by the decrease in fluvial nutrient supply and/or a change in nutrient content of the upwelled waters.

(4) Near 2.5 Ma, accumulation rates of (marine) organic carbon distinctly increased at Site 660. During the last 2.5 Ma, the glacial sediments are carbonate-lean and enriched in (marine) organic carbon, probably caused by the influence of a carbonate-dissolving and oxygen-poor deep water mass.

144

7. Accumulation of organic carbon in (anoxic) marginal seas: I. The Mediterranean Sea (DSDP-Leg 42 and ODP-Leg 107).

From DSDP/ODP Legs 13, 42A, and 107 as well as from numerous piston core data it is well known that episodically organic-carbon-rich sediments ("sapropels") were formed in the Mediterranean Sea during Plio-Pleistocene times (e.g., Olaussen, 1961; Thunell et al., 1977; Kidd et al., 1978; Sigl et al., 1978; Calvert et al., 1987; ten Haven et al., 1987). These sapropels do not occur in the whole Mediterranean Sea but are mainly restricted to the eastern basins (Fig. 101; Stanley, 1978; Thunell et al., 1984).

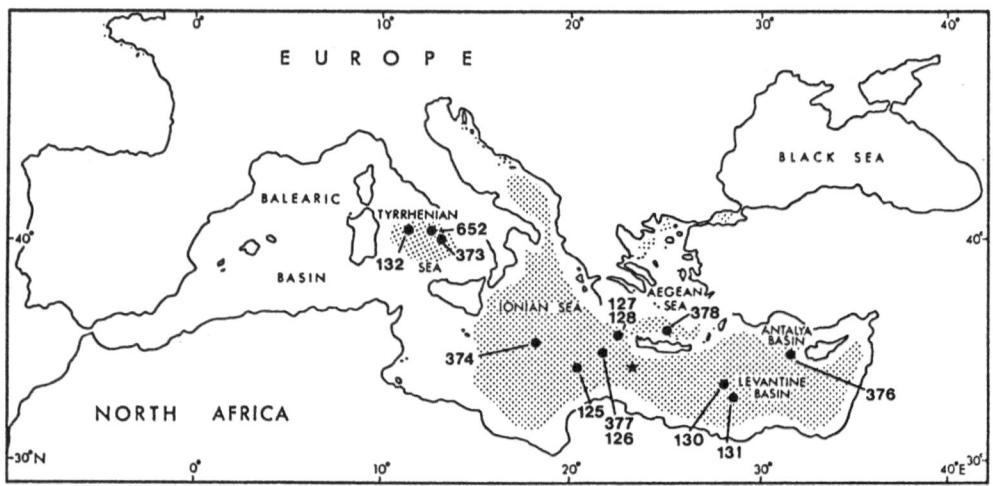

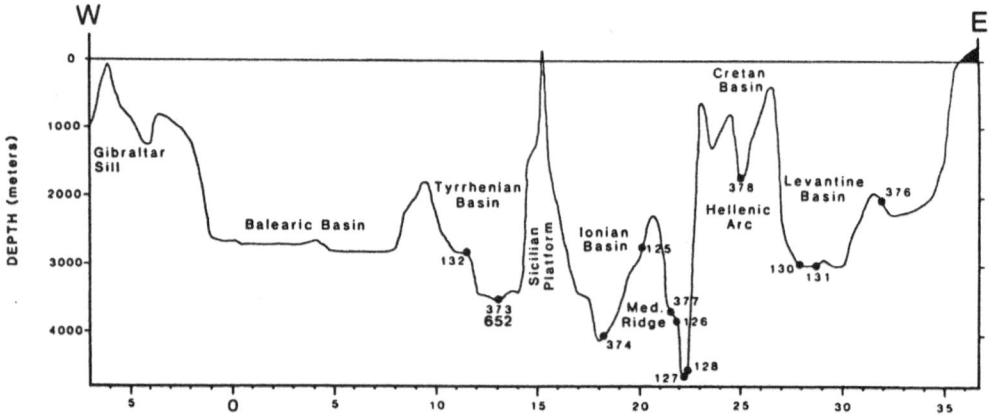

Fig. 101: A. Location map of DSDP/ODP-Sites and distribution of sapropels (after Stanley, 1978; Thunell et al., 1984; Kastens et al., 1987). Asterix marks position of Core RC 9-181 (see Fig. 102). B. W-E-Profile through the Mediterranean Sea with site positions where sapropels have been recorded (from Thunell et al., 1984; suppl.).

During ODP-Leg 107, well developed Pleistocene sapropels were also drilled at Site 652 in the central Tyrrhenian Sea (Fig. 101, Site 652; Kastens, Mascle et al., 1987). Most of the sapropels are associated with climatic warming during interglacials (Fig. 102), indicating that their formation may be climatically controlled (Thunell et al., 1984; Thunell, 1986). Since the sedimentological, micropaleontological, and geochemical characteristics of the Mediterranean sapropels from different time intervals are different, different mechanisms causing the formation of these organic-carbon-rich sediments are likely. These mechanisms, together with new data from DSDP-Sites 374, 376, and 378 (Fig. 101), are briefly presented in the following, and the Mediterranean model will be compared with the results from the Sea of Japan in order to help understanding the formation of organic-carbon-rich sediments in the Sea of Japan (Chapter 8.).

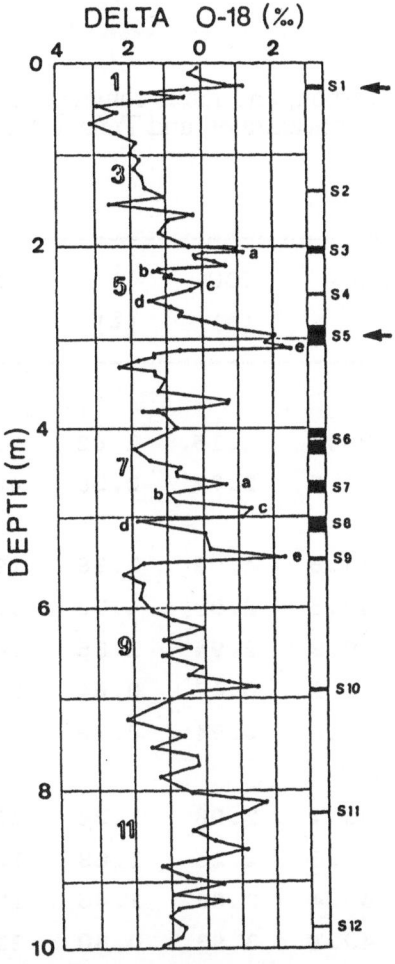

Fig. 102: Oxygen isotope startigraphy (determined on *G. ruber*) and occurrence of sapropel at Core RC 9-181 (from Vergnaud-Grazzini et al., 1977). Marked sapropels S-1 and S-5 are discussed in the text. For core position see Figure 101.

The late Pliocene/Pleistocene Mediterranean sapropels are characterized by high organic carbon contents of up to 10 % (Tab. 8, Fig. 103; e.g., Kidd et al. 1978; Sigl et al. 1978). According to the data from Rock-Eval pyrolysis performed on bulk sediments and kerogen concentrates (Figs. 104 and 105), C/N ratios (Fig. 104), and kerogen microscopy (Stein, unpubl.), the investigated sapropels from Sites 374, 376 and 378 are generally enriched in marine organic carbon, however, with a significant amount of terrigenous fraction. In the single sapropel from Site 374, terrigenous compounds may be the dominant fraction of the organic matter, as indicated by a very low hydrogen index and a very high C/N ratio (Tab. 8). Major to dominant amounts of marine organic carbon in most of the Mediterranean sapropels is also supported by several other detailed studies on the composition of organic matter of these sapropels (e.g., Sutherland et al., 1984; Poutanen and Morris, 1985; ten Haven et al., 1987).

Tab. 8: Carbonate, organic carbon, and sulfur contents, C/N ratios, hydrogen indices, atomic H/C ratios of kerogen concentrates, and Tmax values of sediment samples from DSDP-Sites 374, 376, and 380.

Sample	Depth (m)	$CaCO_3$ (%)	TOC (%)	S (%)	C/N	HI (mgHC/gC)	H/C	T_{max} (°C)
374								
05-2-040	298.91	26.1	1.15	1.82	23	380	1.1	428
06-2-042	332.43	29.5	2.77	0.58	40	73	0.8	429
376								
02-2-123	10.24	43.6	4.08	1.38	19	345	1.2	414
02-3-006	10.57	49.5	2.98	4.72	10	580	1.0	406
03-2-035	18.86	49.9	2.94	1.86	14	321	0.8	416
05-2-138	38.89	48.1	0.47	0.64	24	123	-	-
06-4-092	50.93	60.8	0.94	0.36	31	111	-	-
378								
01-2-015	85.66	25.4	2.08	1.34	21	235	1.3	423
03-3-113	107.14	34.8	4.25	1.68	16	480	1.1	422
06-3-100	145.00	55.4	2.88	1.46	13	672	-	423
08-1-112	218.13	40.1	3.93	0.80	21	385	-	404
11-2-031	304.32	36.5	1.41	2.03	23	125	1.1	-

The presence of sediments characterized by high amounts of marine organic carbon in the marine environment requires unusual depositional conditions (cf., Chapter 2). To explain the formation of the late Pliocene/Pleistocene sapropels in the Mediterranean Sea, both the "stagnation model" and the "high-productivity model" (cf., Fig. 132) have been used (e.g., Olaussen, 1961; Thunell et al., 1977; Schrader and Matherne, 1981; Calvert, 1983; Sutherland et al., 1984; Thunell and Williams, 1989).

Probably both the stagnation and the productivity model (or a combination of both) are useful to explain the formation of sapropels of different age and in different areas, as will be shown in two examples. In the area south to southeast of Crete, sapropel S-5 (lowermost isotopic stage 5; Fig. 102) is distinctly enriched in well-preserved marine diatoms whereas in the younger sapropels S1 and S2 diatoms are absent (Schrader and Matherne, 1981). This suggests that during the formation of sapropel S5 surface-water productivity was high, but during times of formation of sapropels S1 and S2 an anoxic environment causing increased preservation of marine organic matter, was more important in that area. Both organic-carbon- and biogenic-opal-rich as well as organic-carbon-rich and biogenic-opal-poor intervals are also recorded in the sediments from the Sea of Japan (cf., Chapter 8).

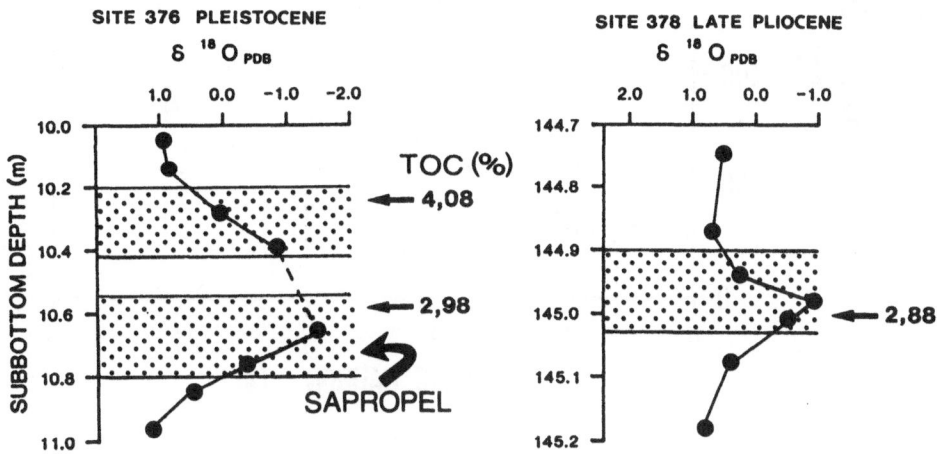

Fig. 103: Oxygen iosotope composition of *Globigerinoides ruber* (from Thunell, 1986) and total organic carbon contents of sapropels at DSDP-Sites 376 and 378 (cf., Tab. 8).

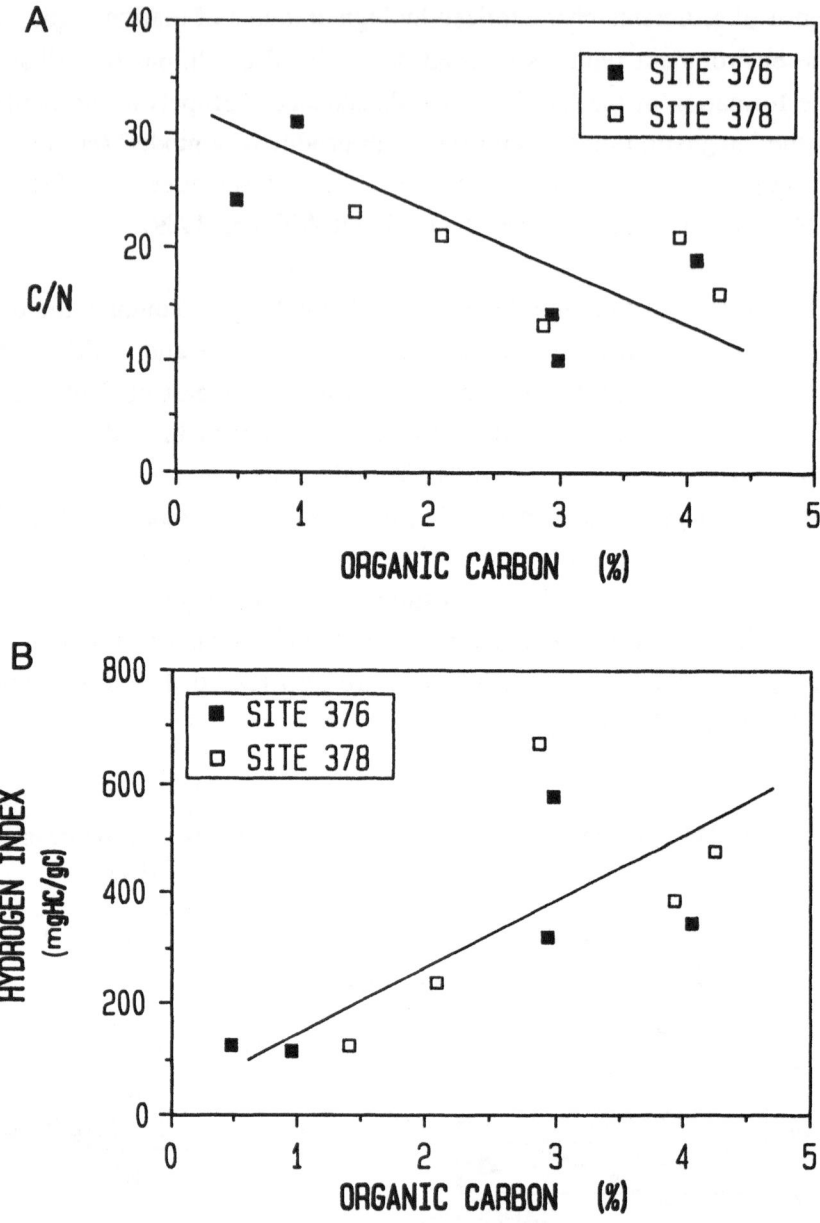

Fig. 104: Correlation between C/N ratios and total organic carbon contents (A) and between hydrogen indices and total organic carbon contents (B) of samples from DSDP-Sites 376 and 378.

A second example is from sapropel S-1 deposited in the Outer Helenic Ridge area, close to DSDP-Site 125 (Fig. 101; Sutherland et al., 1984). High amounts of marine organic carbon indicated by a mean stable carbon isotopic value of -20.6 $^O/_{oo}$ and a C/N ratio around 10, increased barium values of up to 1000 ppm, and distinctly increased sedimentation rates of about 23 cm/1000yrs point to increased surface-water productivity controlling sapropel formation (Sutherland et al., 1984; cf., Chapter 3).

Based on a detailed study on oxygen isotopes of planktonic foraminifera from sediments of the last glacial/interglacial cycle, Thunell and Williams (1989) presented a model to explain the formation of the most recent sapropel S1 (7000-9000 years BP; Sutherland et al., 1984), combining stagnation and high-productivity mechanisms (Fig. 106): During the last glacial maximum, highly-saline surface and deep waters (with salinities probably higher than those of today) formed in the eastern Mediterranean basins because evaporation exceeded precipitation and runoff under more arid climatic conditions in the circum-Mediterranean area. This resulted in an anti-estuarine water exchange with the North Atlantic (Fig. 106, situation a).

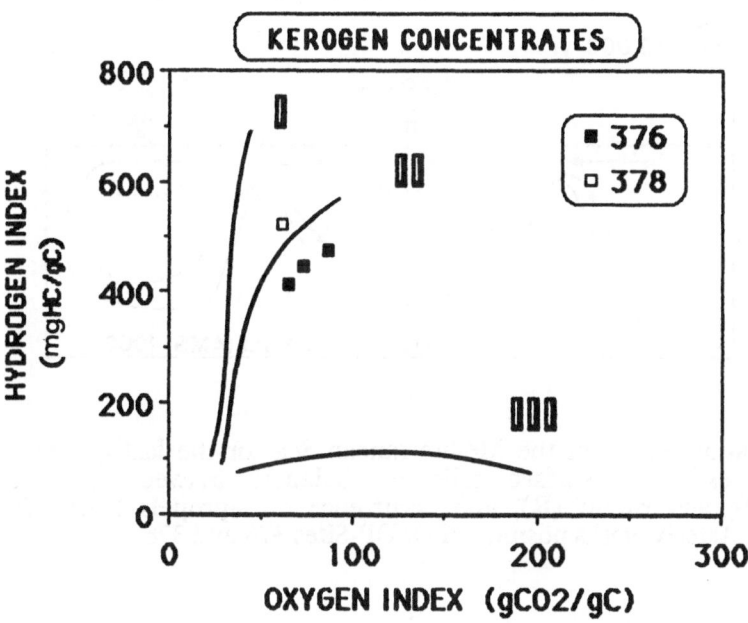

Fig. 105: Hydrogen and oxygen indices determined on kerogen concentrates of selected samples from Sites 376 and 378.

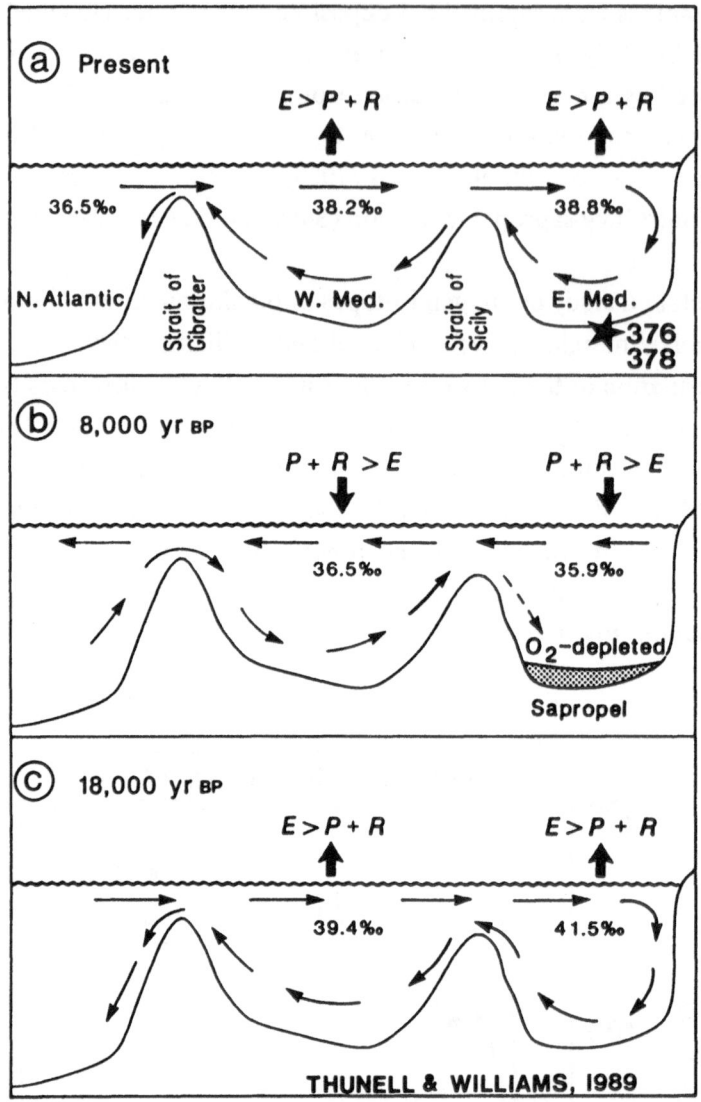

Fig. 106: Circulation model of the Mediterranean Sea for the last glacial/interglacial cycle, showing estimated surface salinities, balances between evaporation (E), precipitation (P), and runoff (R), and occurrence of sapropels (from Thunell and Williams, 1989). Asterix marks position of DSDP-Sites 376 and 378.

With the onset of deglaciation near 12500 years BP, the climate became more humid and precipitation and runoff increased (e.g., Rossignol-Strick et al., 1982) and exceeded the rate of evaporation, resulting in an estuarine water exchange with the Atlantic (Fig. 106, situation b). The low-salinity surface water and the very saline bottom water (formed during glacial times) allowed the establishment of a very stable density stratification inducing euxinic conditions on the sea floor, favorable for increased preservation of organic matter. In addition, the inflow of deep-water from the Atlantic would have caused the eastern Mediterranean to become a nutrient trap (Thunell and Williams, 1989). This as well as the fluvial nutrient supply may have enhanced surface-water productivity, too (Calvert, 1983). This means that in addition to deep-water stagnation, an increased production rate of marine organic matter may have controlled the formation of sapropels. The absence or presence of diatoms in these sapropels (see above), the relationships between both organic carbon content and sedimentation rate and organic carbon content and (pyritic) sulphur content (Leventhal, 1983; Stein, 1990) as well as the barium content (Schmitz, 1987; Bishop, 1989) may help to estimate the importance of surface-water productivity for sapropel formation.

Since most of the older sapropels from the eastern Mediterranean Sea also occur in intervals marked by distinctly depleted oxygen isotope values of planktonic foraminifera (Fig. 103), reduced surface-water salinity probably prevailed during their formation (Thunell et al., 1977; Vergnaud-Grazzini et al., 1977; Thunell, 1986). Furthermore, the recognition of lower Pleistocene sapropels in the central Tyrrhenian Sea at Site 652 (Kastens, Mascle et al., 1987) indicates that periods of enhanced preservation of organic matter and/or increased surface-water productivity should have occasionally occurred also in this part of the Mediterranean Sea. Further sedimentological, organic and inorganic geochemical, and paleontological work on Mediterranean sapropels is necessary to reconstruct the process of sapropel formation through space and time in more detail.

8. Accumulation of organic carbon in (anoxic) marginals seas: II. The Sea of Japan (ODP-Leg 128)
(R. STEIN and the SHIPBOARD SCIENTIFIC PARTY)

8.1. Introduction

ODP-Legs 127 and 128 were planned to penetrate sediments that would allow the reconstruction of the tectonic, depositional, and paleoceanographic evolution of the Japan Sea (Ingle, Suyehiro, et al., 1990; Pisciotto, Tamaki, et al., 1990). During Leg 128, two Sites (798 and 799; Fig. 107) were drilled on Oki Ridge and in the Kita-Yamato-Trough, respectively (Fig. 108). Furthermore, Site 794 located in the northern Yamato Basin (Fig. 107), was selected for both downhole geophysical experiments and for deeper drilling into basement rocks initially cored during Leg 127 (Ingle, Suyehiro et al., 1990).

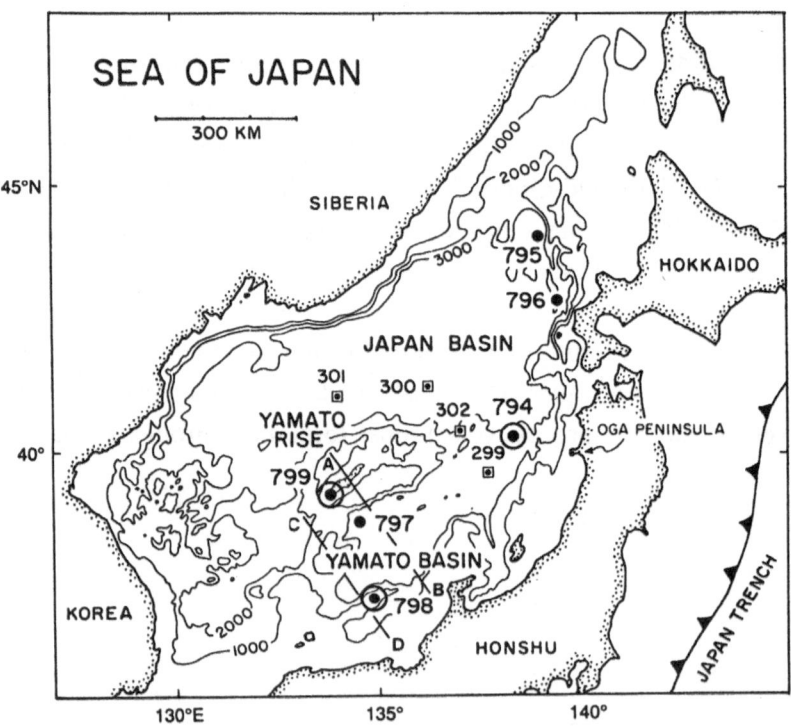

Fig. 107: Location map of DSDP-Leg 31 Sites (299 - 302), ODP-Leg 127 Sites (794 - 797), and ODP-Leg 128 Sites (798 and 799). Lines A-B and C-D mark seismic profiles of Figure 108.

The primary objective at Site 798 was to obtain a Miocene to Quaternary carbonate-rich sequence, which would provide a high-resolution paleontologic, isotopic, and sedimentologic record. Of major interest were reconstructions of (1) surface- and intermediate-water history including changes in surface-water productivity, oxygen content of water masses, and CCD, and (2) changes from anoxic to oxic conditions known to have occurred during Miocene to Pliocene times (Ingle, 1975; Matoba, 1984), and (3) faunal and sedimentary responses to the tectonic evolution of the Japan Sea (Ingle, Suyehiro et al., 1990).

The main objective at Site 799 was to reconstruct the depositional and tectonic history of the Kita-Yamato-Trough as typical environment for massive sulfide mineralization in a rifted continental-arc setting (Ingle, Suyehiro et al., 1990).

At Sites 798 and 799, organic-carbon-rich sediments were deposited. Thus, these sites are of major interest for studies on mechanisms controlling the formation of organic-carbon-rich sediments in a marginal-sea environment which occasionally was restricted from the normal open-ocean environment (cf., Ingle, 1975; Matoba, 1984).

The data presented and discussed here are mainly shipboard data produced onboard "JOIDES RESOLUTION" during Leg 128, supplemented by first results of shore-based analyses. Thus, the interpretations are still preliminary. Further detailed sedimentological, geochemical, and paleontological studies are in progress by the participants of Leg 128, the results of which will hopefully allow more complete and precise interpretations in the near future.

8.2. Recent and sub-recent environment in the Japan Sea

The Japan Sea is a semi-closed back-arc basin with maximum water depths of about 3600 m (Fig. 107). It is separated into several major basins, i.e., Japan Basin, Yamato Basin, Tartary Trough, and Tsushima Basin by topographic highs which are continental fragments or tectonic ridges (Figs. 107 and 108). The ages of formation of these basins were estimated from sediment stratigraphy, basement depth, and heat flow data to vary between 30 and 10 Ma (Tamaki, 1985, 1988). The connections to marginal seas in the North and in the South, and to the western Pacific are restricted to shallow straits, the Tsushima Strait (140 m water depth), the Tsugaru Strait (130 m water depth), the Soya

154

Strait (55 m water depth), and the Tartary Strait (15 m water depth) (Fig. 109; Ingle, 1975; Matoba, 1984).

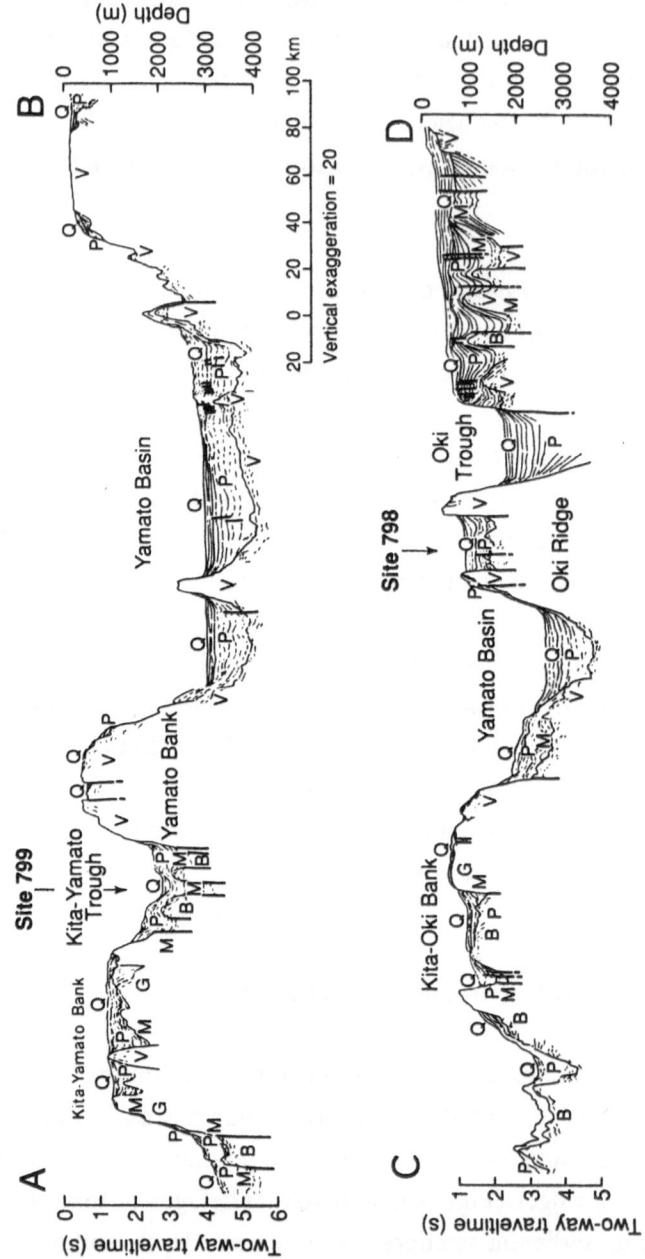

Fig. 108: Seismic profile interpretations of profiles through the areas of Sites 798 and 799 (from Tamaki, 1988; Ingle, Suyehiro et al., 1990). For positions of profiles see Figure 107.

The modern surface-water circulation in the Japan Sea is dominated by the inflow of warm and highly-saline water through the Tsushima Strait (the Tsushima Current, a branch of the Kuroshio Current), the inflow of cold water of low salinity from the north (the Liman Current), and the outflow of water masses through the Tsugaru Strait and the Soya Strait into the western Pacific (Figs. 109 and 110). The modern deep water of the Japan Sea ("Japan-Sea Proper Water") forms in the northwestern coastal areas during winter times due to cooling and freezing of surface water. The resulting water masses are of higher density and sinks to the sea floor (Fig. 110). This intense thermohaline circulation results in a sufficient ventilation of the deep-water sphere; dissolved oxygen content at all depths exceeds 5 ml/l (Matoba, 1984). Because of these oxidizing oceanographic conditions, organic-carbon contents of surface sediments are unusually low (Ingle, 1975).

The environmental conditions, however, must have changed dramatically during late Pleistocene to Holocene times as indicated by piston core data. During glacial intervals, i.e., times of distinctly lowered sea level, the Japan Sea was probably completely isolated from the Pacific Ocean. The expansion and duration of sea ice coverage, loss of exchange with the Pacific Ocean, and sluggish bottom-water circulation may have caused anoxic deep-water conditions allowing the formation of framboidal pyrite and organic-carbon-rich sediments (Kobayashi and Nomura, 1972; Ujiie and Ichikura, 1973; Ingle, 1975; Oba, 1983; Oba et al., 1990).

8.3. Geological setting and sediments at ODP-Sites

Site 798 is located in the southeastern Japan Sea, about 160 km north of the western coast of Honshu in a small sediment-filled graben on top of Oki Ridge (Figs. 107 and 108, Tab. 9). This shallow-water position well above the modern CCD (1500 -2200 m; Ujiie and Ichikura, 1973) and on a structurally isolated high was chosen to obtain a carbonate-rich sequence undiluted by coarse-grained gravity-flow sediments common to basinal areas (Ingle, Suyehiro et al., 1990).

At Site 798, a 517 m thick pelagic to hemipelagic sediment sequence of late early Pliocene to Holocene age was drilled (Ingle, Suyehiro et al., 1990). Major lithologies are diatomaceous ooze, diatomaceous clay, silty clay, clay, and siliceous claystone, with common occurrence of foraminifera and calcareous nannofossils in the upper 220 m of the sequence.

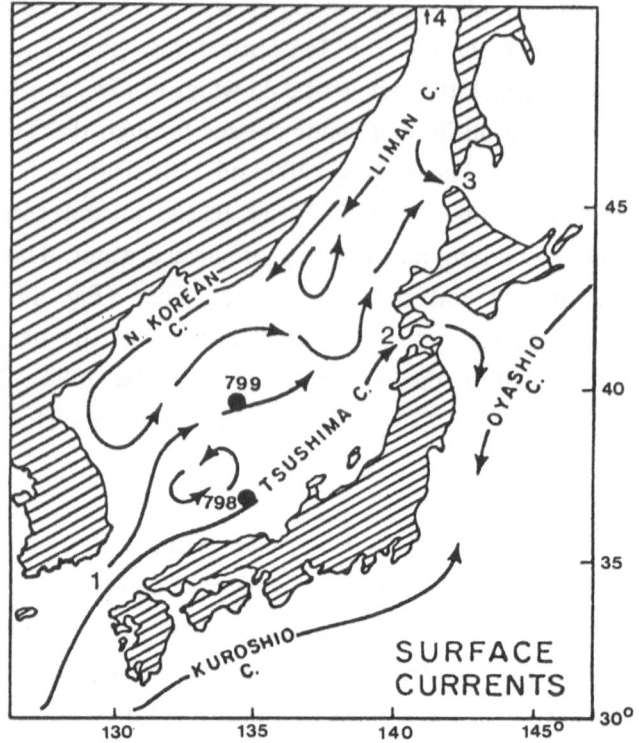

Fig. 109: Surface-water currents in the Japan Sea (from Ingle, 1975). Numbers indicate Tsushima Strait (1), Tsugaru Strait (2), Soya Strait (3), and Tartary Strait (4).

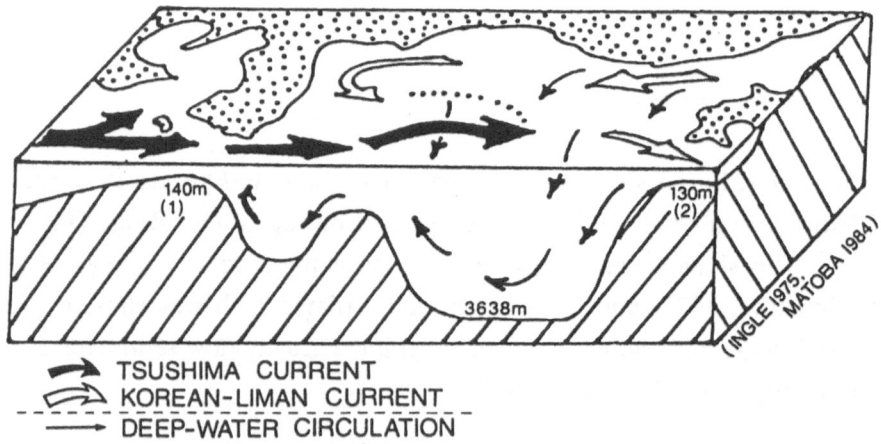

Fig. 110: Scheme of modern surface- and deep-water circulation in the Japan Sea (from Ingle, Suyehiro, et al., 1990; after Ingle, 1975 and Matoba, 1984). Large black arrows indicate Tsushima Current; open arrows indicate Korean-Liman Current; small black arrows indicate deep- and intermediate-water circulation. Dotted line marks front between cold-low salinity Liman-Current water and warm-high salinity Tsushima-Current water. (1) and (2) indicate Tsushima Strait and Tsugaru Strait, respectively.

This upper part is furthermore characterized by rythmic changes between dark laminated, diatom- and organic-carbon-rich intervals and light, homogeneous to intensely bioturbated, clay-rich intervals. Volcanic ash layers are common, with a maximum occurrence between 0.9 and 0.3 Ma (Ingle, Suyehiro et al., 1990).

Based on shipboard paleomagnetic and microfossil data (Burckle et al. and Hamano and Krumsiek in Ingle, Suyehiro et al., 1990), average sedimentation rates of the entire sequence at Site 798 are about 12 cm/ky (Fig. 111). Intervals of reduced sedimentation probably occur between 210 and 190 mbsf (i.e., 1.9 to 1.7 Ma) and between 90 and 50 mbsf (i.e., 0.9 to 0.5 Ma) (Fig. 111).

Tab. 9: Geographic locations and water depths of ODP-Sites 798 and 799

ODP-Site	Position	Water depth
798	37.04° N, 134.80° E	903.1 m
799	39.22° N, 133.87° E	2073.0 m

Site 799 is located in the Kita-Yamato-Trough, a narrow sediment-filled graben within the much larger Yamato Rise in the south-central Japan Sea (Figs. 107 and 108, Tab. 9). At this Site, a 1084 m thick hemipelagic sediment sequence of lower Miocene to Holocene age was drilled (Ingle, Suyehiro et al., 1990). Major lithologies are diatomaceous ooze/clay, silty clay, clay, siliceous claystone, and porcellanite with common occurrence of turbidites and slumps and authigenic carbonates. Ash layers occur throughout the entire sediment sequence with a maximum occurrence in the upper 100 m. The middle and late Miocene deposits in the Kita-Yamato-Trough were dominated by laminated, organic-matter-rich siliceous claystones with authigenic dolomites. Very similar deposits of same age and characterized by rhythmical alternation of dark-colored more clayey and light-colored more siliceous layers were also recorded onshore in northern Japan (e.g., the Onnagawa Formation and the Odoji Formation; Tada et al., 1986; Iijima et al., 1988) and the coastal areas of California (i.e., the Monterey Formation; Ingle, 1981; Isaacs et al., 1983). Diagenetic opal-A/opal-CT/authigenic quartz boundaries are well defined in this interval (Ingle, Suyehiro et al., 1990; cf., Fig. 124).

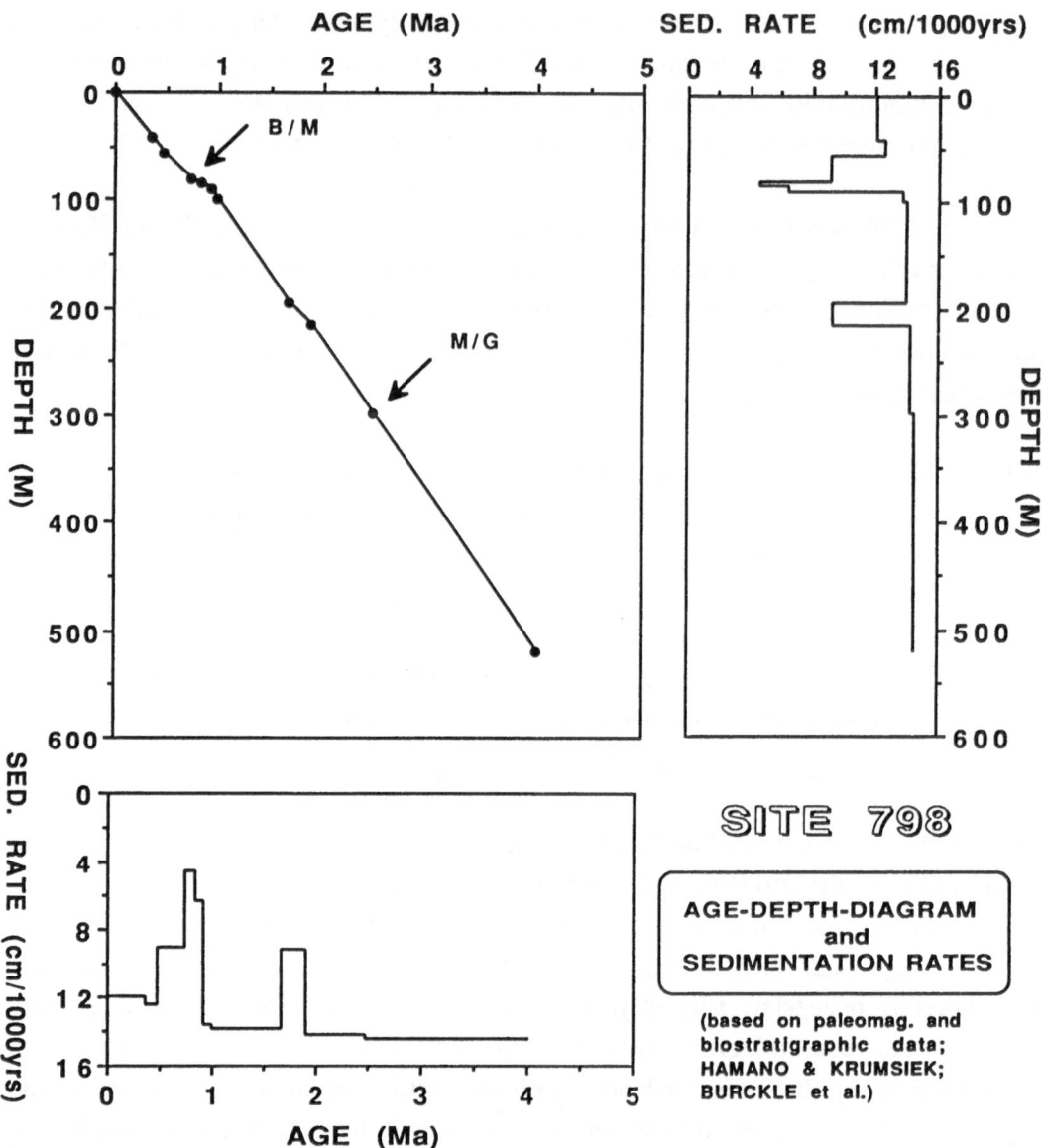

Fig. 111: Age-depth diagram and sedimentation rates at Site 798, based on shipboard paleomagnetic and biostratigraphic data (after Burckle et al. and Hamano and Krumsiek in Ingle, Suyehiro, et al., 1990).

In the upper 468 m (i.e., the last about 6.5 Ma) of Site 799, sedimentation rates vary between about 2 cm/ky and 13 cm/ky with an average of approximately 7 cm/ky, as based on shipboard paleomagnetic data (Hamano and Krumsiek in Ingle, Suyehiro et al., 1990). For the lower part of the sequence of Site 799 (i.e., Hole 799B), no sedimentation rates were calculated because weak stratigraphic data are only available until now.

8.4. Preliminary Results

8.4.1. Quantity of organic carbon and carbonate content

The entire sequence of Pliocene to Quaternary sediments from Site 798 is characterized by high organic carbon values ranging between about 1 and 5 % (Fig. 112). Maximum values of 3 to 5 % are concentrated in the intervals between 80 and 120 mbsf (i.e., between 0.8 and 1.2 Ma) and below 500 mbsf (i.e., between 3.8 and 4 Ma). Carbonate contents are very low (less than 4 %) in the lower part of the sequence, starting to increase at about 220 mbsf (i.e., 1.9 Ma) and reaching maximum values of more than 25 % between 30 and 80 mbsf (i.e., between 0.3 and 0.8 Ma) (Fig. 112). High-amplitude variations in both organic carbon and carbonate contents are typical of the upper 140 m of the sediment sequence at Site 798.

At Site 799, organic carbon values are also high and vary between 0.5 and 6 % in the Miocene to Quaternary sediment sequence (Fig. 113). Maximum organic carbon values of 3 to 6 % are concentrated in the late middle to late Miocene, whereas in the Plio-Pleistocene interval lower values of 0.5 to 1.8 % are typical. However, single organic-carbon spikes of 3 to 5 % also occur during the latter time interval (Fig. 113). In general, carbonate contents at Site 799 are low ranging between 0 and 10 % (Fig. 113). The single spikes with high carbonate values of 30 to more than 70 % are from thin foraminifera-rich beds and authigenic carbonate layers or nodules (cf., Ingle, Suyehiro, et al., 1990).

To interpret the organic carbon and carbonate data in terms of changes in flux of both variables, percentage values were transformed into accumulation rates (cf., Chapter 4). Accumulation rates have not been calculated for Site 799 because of major occurrence of turbidites and slumps (Ingle, Suyehiro et al., 1990).

160

At Site 798, accumulation rates of organic carbon generally vary between 0.1 and 0.3 gC cm^{-2} ky^{-1} (Fig. 114). Between 0.9 and 1.3 Ma and between 3.8 and 4 Ma, however, higher values of more than 0.4 gC cm^{-2} ky^{-1} have been recorded. Carbonate accumulation rates are low prior to 1.7 Ma (0.1 to 0.6 gC cm^{-2} ky^{-1}; Fig. 114). The last 1.7 Ma are characterized by high-amplitude variations between 0.1 and 2.5 gC cm^{-2} ky^{-1}).

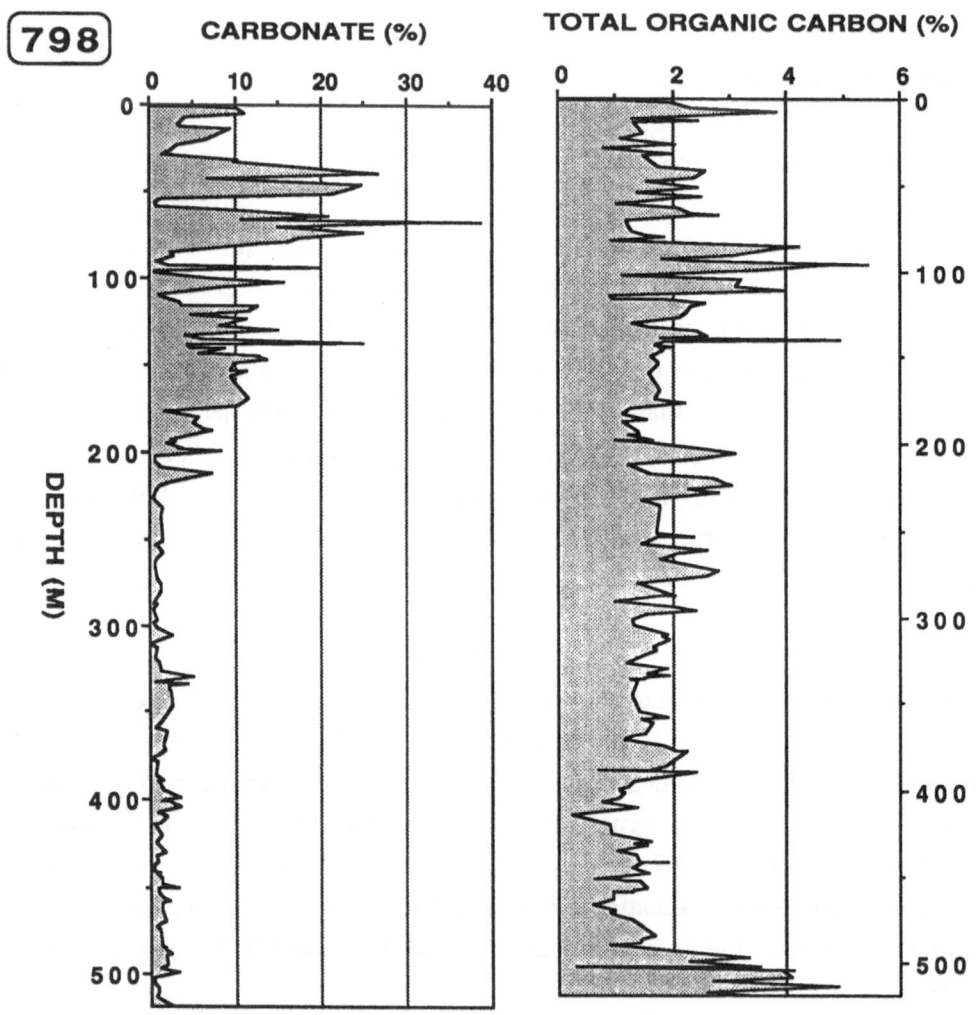

Fig. 112: Carbonate and organic carbon contents vs. depth at Site 798.

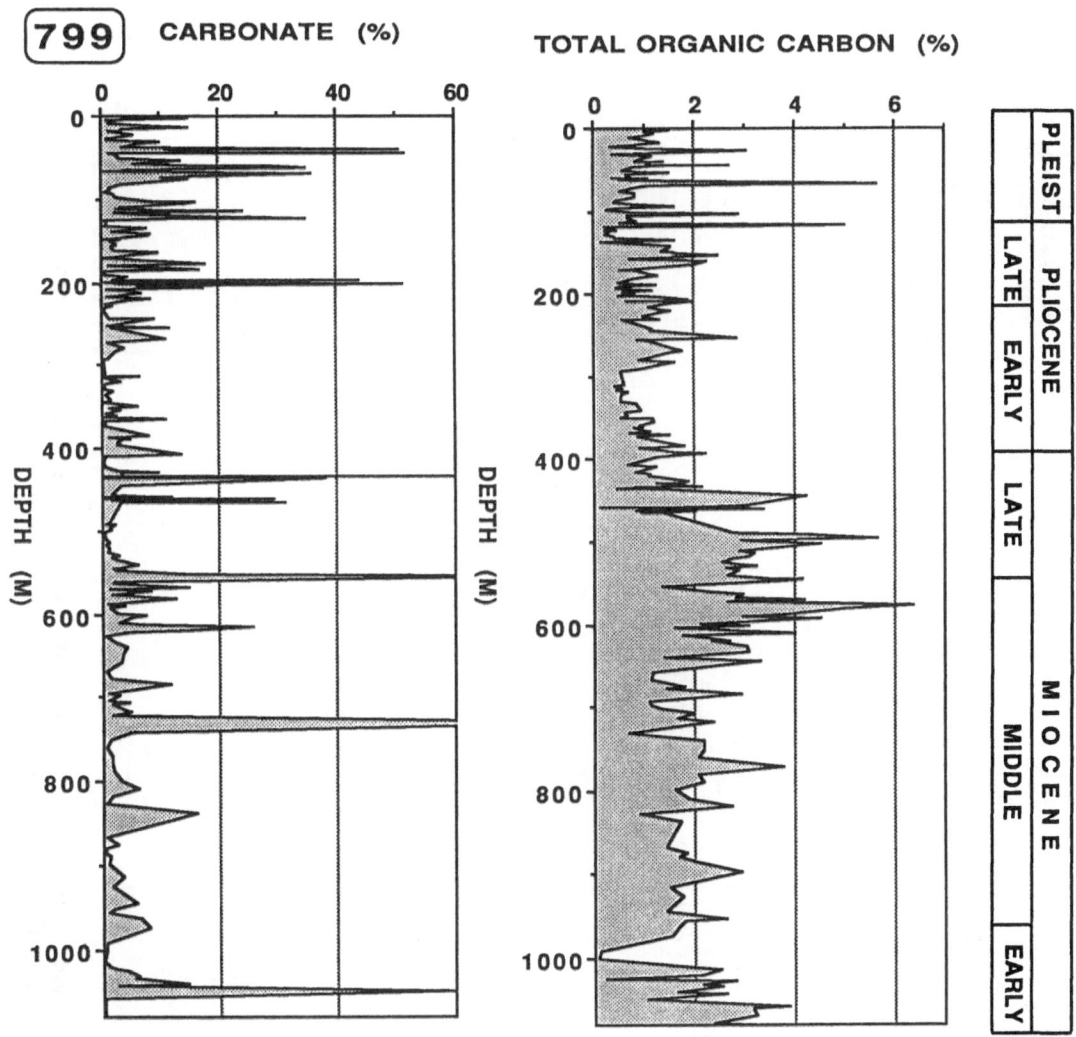

Fig. 113: Carbonate and organic carbon contents vs. depth at Site 799.

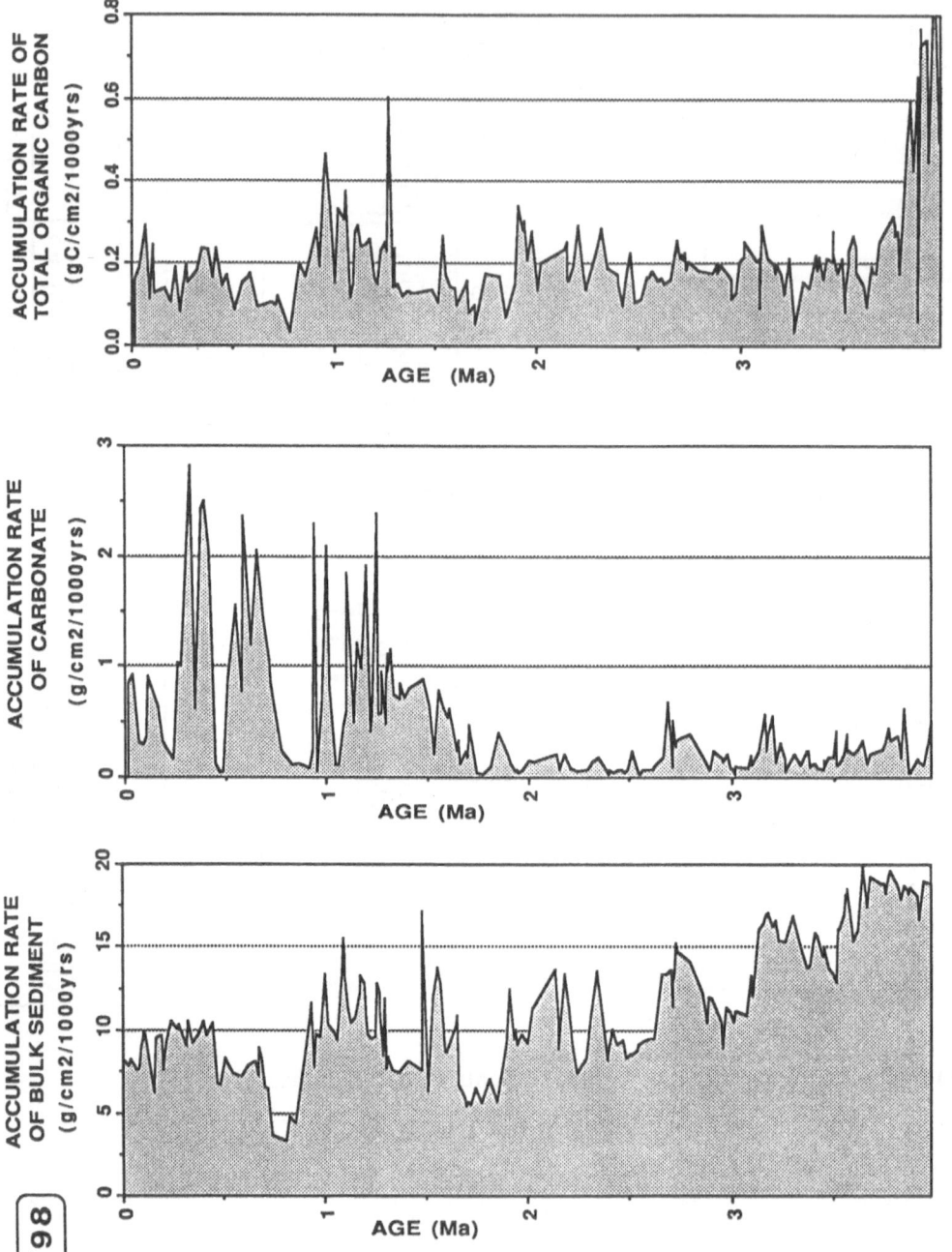

Fig. 114: Accumulation rates of bulk sediment fraction, carbonate, and organic carbon vs. age at Site 798, based on shipboard stratigraphy.

Comparing the accumulation rates of both organic carbon and carbonate with bulk accumulation rates which vary between 3 and 20 gC cm^{-2} ky^{-1} (Fig. 114), it is obvious that the dominant sediment fractions at Site 798 are non-organic-carbon and non-carbonate components, i.e., biogenic opal and siliciclastics (Ingle, Suyehiro et al., 1990).

8.4.2. Composition of organic matter

Data on organic carbon composition available till now are mainly results from elemental analyses and Rock-Eval pyrolysis. Kerogen microscopy data, carbon stable isotopes, and biomarker data which would allow a more detailed characterization of the organic carbon fraction, are in progress (e.g., Kettler et al., in prep.; Stein and Stax, in prep.). Thus, the classification of the organic carbon is still very preliminary.

Organic carbon/nitrogen (C/N) ratios

At Site 798, most of the C/N ratios vary between 5 and 12 (Figs. 115 and 116), suggesting a marine-type of organic matter to be dominant. In several samples from the lower Pleistocene and mid-Pliocene intervals, very high nitrogen contents of 0.3 to 0.5 % and organic carbon contents of 0.5 to 2.5 % have been determined resulting in very low C/N ratios of 2 to 5 (Fig. 115, A and C). They occur as single spikes throughout these intervals (Fig. 116), suggesting different primary composition of organic matter (or increased content of inorganic nitrogen) rather than diagenetic alteration. Near the bottom of Site 798 (i.e., in the late early Pliocene), maximum organic-carbon contents coincide with increased C/N ratios of about 15 (Fig. 116) which may indicate some higher amounts of terrigenous organic carbon.

At Site 799, most of the C/N ratios range between 5 and 20, with some single spikes of up to 30 (Figs. 117 and 118). In the Plio-Pleistocene interval, C/N ratios around 10 are dominant (which are similar to those recorded for Site 798), whereas in the Miocene interval higher C/N ratios around 15 are typical, suggesting significant amounts of terrigenous organic carbon deposited during the latter time interval. In general, however, marine organic carbon seems to be the major carbon fraction also at Site 799.

164

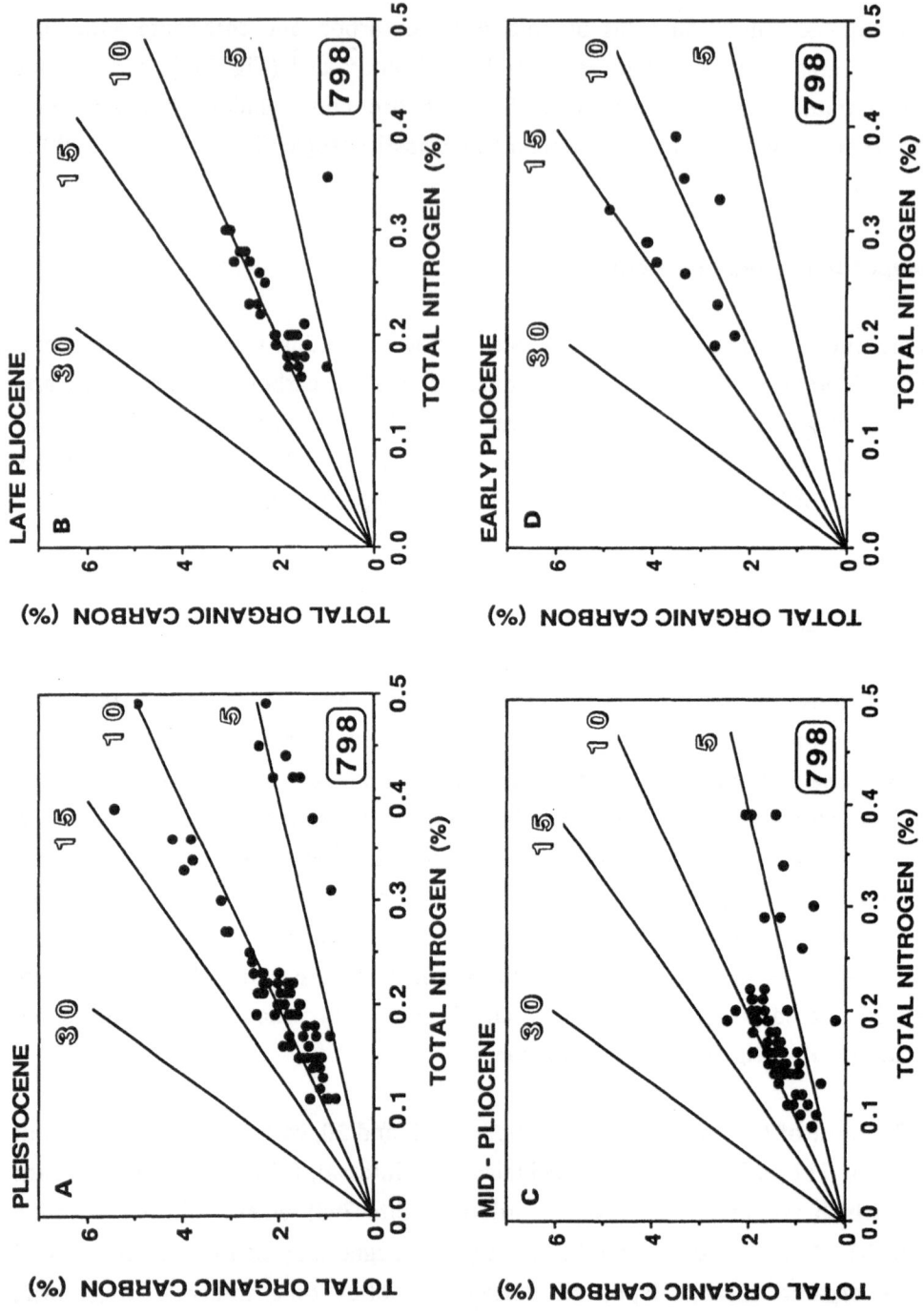

Fig. 115: Total organic carbon vs. total nitrogen ratios at Site 798 for Pleistocene (A), late Pliocene (B), mid-Pliocene (C), and early Pliocene (D) time intervals.

798

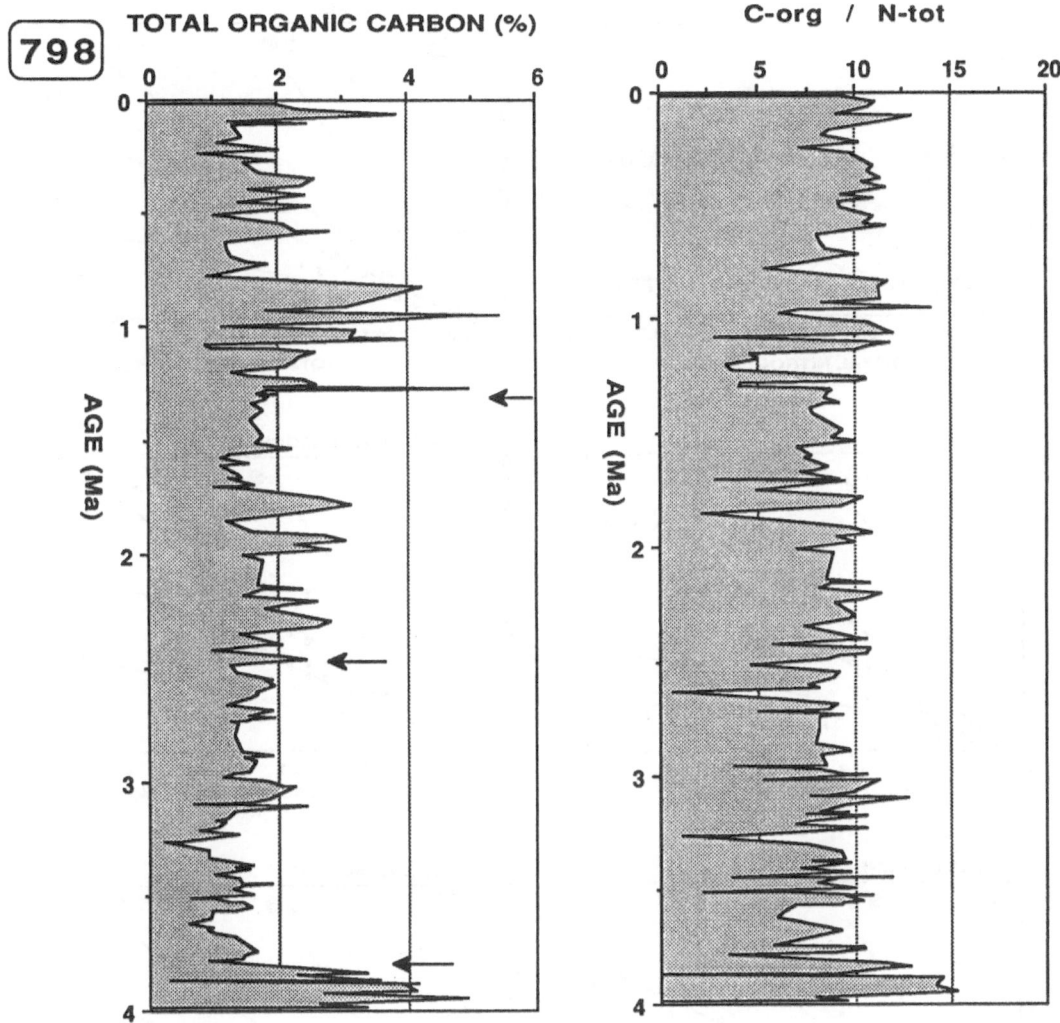

Fig. 116: Total organic carbon contents and C/N ratios vs. age at Site 798.

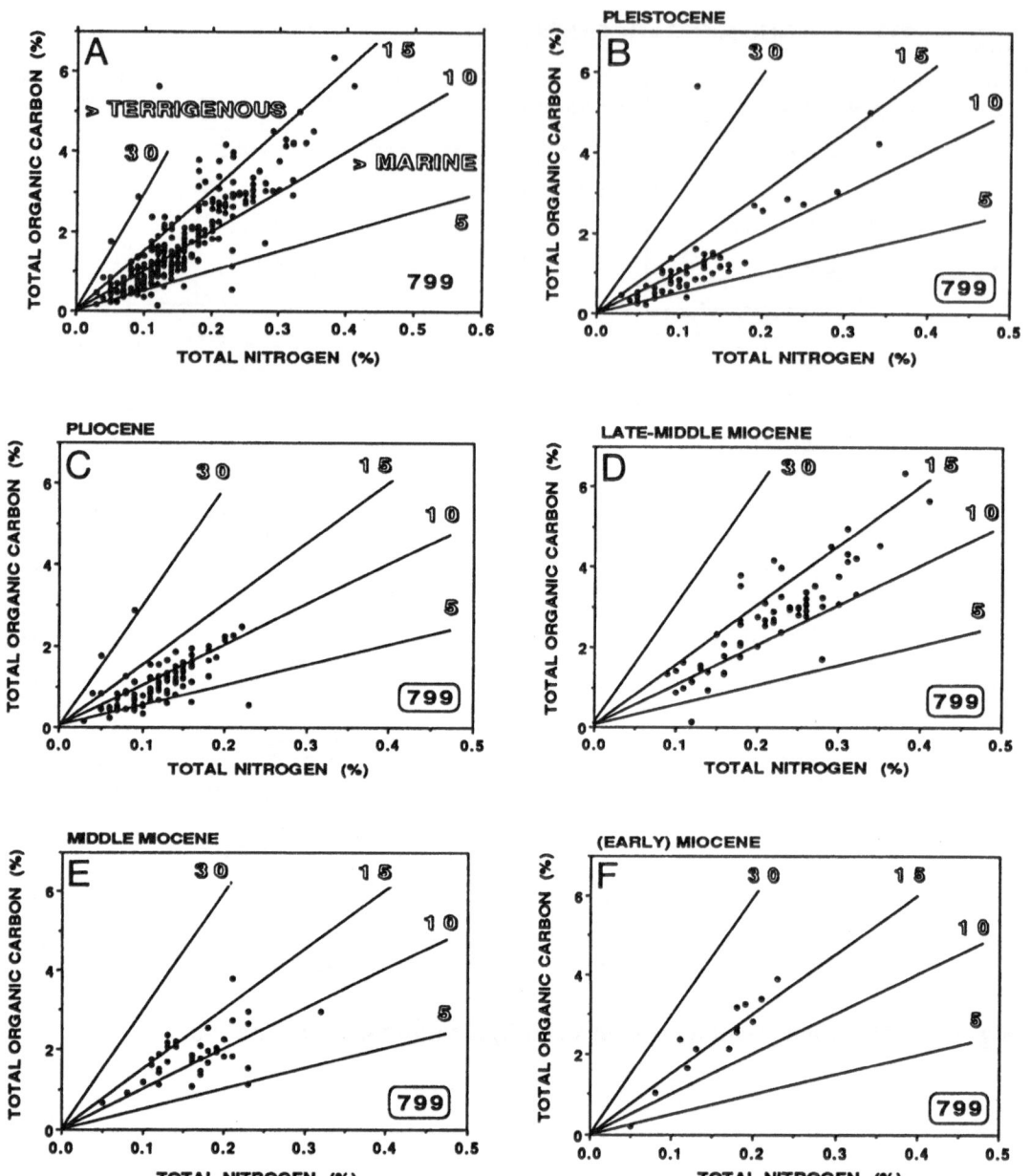

Fig. 117: Total organic carbon vs. total nitrogen at Site 799 for the entire sediment sequence (A) and Pleistocene (B), Pliocene (C), late-middle Miocene (D), middle Miocene (E), and (early) Miocene (F) time intervals.

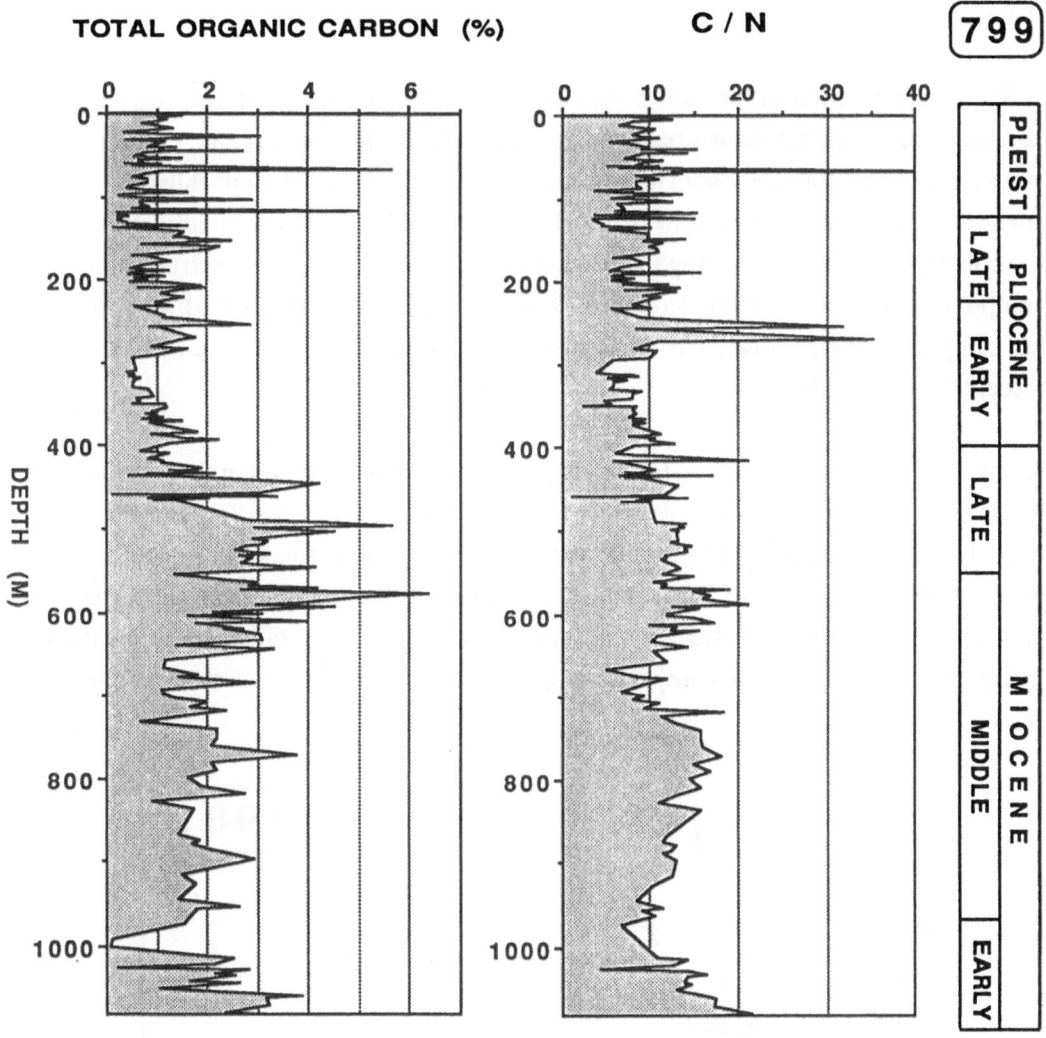

Fig. 118: Total organic carbon contents and C/N ratios vs. depth at Site 799.

Rock-Eval pyrolysis

At Site 798, hydrogen index values vary between 100 and 600 mgHC/gC$_{org}$ (Fig. 119, Tab. 10), indicating a mixture between marine and terrigenous organic carbon, with a dominance of the marine proportion in most of the samples. The unusually high oxygen index values measured on samples from Hole 798A may result from significant amounts of labile organic matter (e.g., sugars, proteins) still present in the immature sediments of the upper about 100 m of the sequence (Tab. 10; cf., Stein et al., 1989b). In carbonate-rich samples, CO$_2$ generated from carbonate during pyrolysis may have also caused increased oxygen indices.

At Site 799, first results of Rock-Eval pyrolysis display hydrogen and oxygen index values similar to those recorded for Hole 798B (Figs. 119 and 120), suggesting that the marine proportion is dominant in the organic carbon fraction also at this Site. Lower hydrogen index values measured in few samples indicate significant amounts of terrigenous organic carbon. More analyses are necessary before a more detailed interpretation of changes in organic carbon composition throughout the entire sediment sequence is possible.

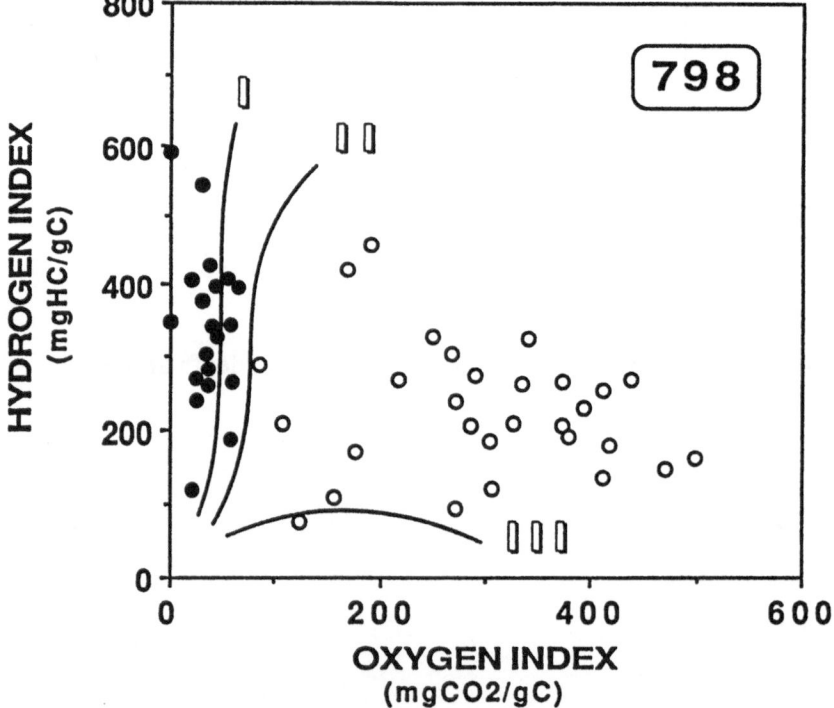

Fig. 119: Results from Rock-Eval pyrolysis: Hydrogen vs. oxygen index values at Site 798A (open circles) and 798B (solid circles).

Tab. 10: Results of Rock-Eval pyrolysis of bulk sediment samples from Sites 798 and 799, performed by R. Stax (Giessen University) onboard "JOIDES RESOLUTION" during Leg 130.

Sample	C_{org} (%)	HI (mgHC/gC)	OI (mgCO$_2$/gC)	T_{max} (°C)
798B				
10-7-025	2.32	196	62	428
10-7-054	4.67	403	54	405
10-7-074	4.61	333	42	415
10-7-075	4.08	349	59	419
10-7-085	4.82	399	46	411
10-7-091	2.94	403	57	414
14-6-065	2.70	268	67	423
19-4-050	2.21	116	15	360
23-7-060	2.71	341	41	419
29-6-035	2.60	314	34	418
35-2-015	1.19	290	31	442
38-6-058	1.65	268	29	421
40-5-031	2.25	430	31	422
46-1-050	1.55	267	19	420
51-4-108	1.55	256	22	422
799A				
03-6-010	1.23	153	37	478
08-6-050	5.64	103	21	414
12-6-020	0.69	42	157	378
16-4-060	1.56	116	81	416
22-6-130	1.03	243	86	411
27-2-025	1.17	248	40	418
41-4-052	1.51	212	27	408
46-6-020	1.36	234	37	410
49-cc-38	4.23	316	20	420
799B				
05-1-014	2.77	241	50	419
06-1-033	5.66	317	28	408
10-4-064	2.91	265	37	417
15-1-025	6.36	438	19	404
19-3-080	2.71	377	55	424
22-1-006	3.31	415	32	422
37-1-055	2.17	493	24	418
40-1-004	2.74	499	24	412
45-1-123	1.42	327	17	413
53-2-000	1.44	297	53	419
63-1-119	1.65	301	82	426
66-2-048	3.25	446	34	417

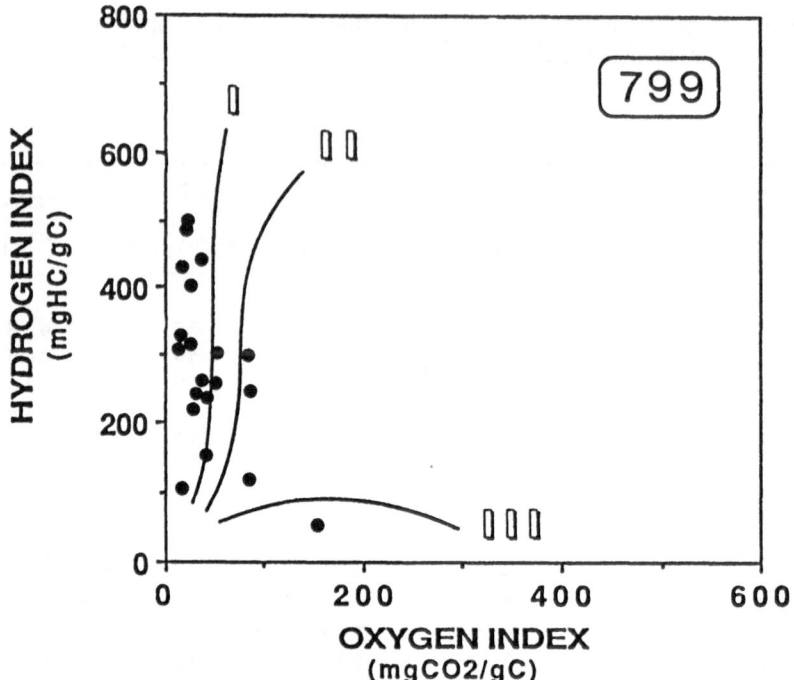

Fig. 120: Results from Rock-Eval pyrolysis: Hydrogen vs. oxygen index values at Site 799.

Comparing the Rock-Eval data from Sites 798 and 799 with the correlation diagram of hydrogen index values and kerogen microscopy data (Fig. 23), rough estimates of marine organic carbon vary between about 20 and 80 % .

Kerogen microscopy
First results of kerogen microscopy performed on sediment samples from Site 798A support the predominance of marine organic matter. The dominant maceral is alginite; well-preserved, yellow to yellowish-green fluorescing bodies of *Tasmanales*-type algae and yellowish-orange to yellow fluorescing laminae (lamalginite) are typical (Stax and Stein, unpubl.).

8.4.3. Maturity of organic matter

According to the hydrogen and oxygen index values, i.e., the position of the data points in the van-Krevelen-type diagrams (Figs. 119 and 120), and the low temperatures of maximum pyrolysis yield (T_{max} values of 378 to 428 ^{o}C; Tab. 10), the organic carbon deposited at Sites 798 and 799 is still immature. Although distinct diagenetic changes indicated by the transition of opal-A/opal-CT/authigenic quartz already occurred in the more than 1000m thick sediment sequence of Site 799 (Ingle, Suyehiro, et al., 1990; Isaacs, et al., 1983; cf., Fig. 124), no increase in maturity with depth is obvious in the Rock-Eval data at this Site. Further data on organic carbon maturity (i.e., vitrinite reflectance and lipid geochemistry data) are necessary for a more detailed reconstruction of the thermal history of the organic carbon fraction.

8.5. Discussion

Using hydrogen index values, the amounts of terrigenous and marine organic carbon have been estimated (cf., Chapter 4) and accumulation rates of both fractions have been calculated (Fig. 121). When interpreting these values, it has to be considered that the estimates are based on (i) shipboard stratigraphy and (ii) a still limited number of hydrogen index values, and thus, (iii) have to be verified by other methods such as kerogen microscopy and lipid geochemistry as well as further stratigraphic data.

8.5.1. Accumulation of terrigenous organic carbon at Site 798 and paleoenvironment

At Site 798, accumulation rates of terrigenous organic carbon vary between 0.02 and 0.2 $gC\ cm^{-2}\ ky^{-1}$, with maxima of 0.1 to 0.2 $gC\ cm^{-2}\ ky^{-1}$ in the uppermost early Pliocene (i.e., near the base of the sediment sequence of Site 798), between 2 and 1 Ma, and near 0.4 Ma (Fig. 121). The oldest sediments recovered at Site 798 and characterized by increased terrigenous organic carbon content are also rich in siliciclastic detritus including glauconite and quartz turbiditic sands (Ingle, Suyehiro et al., 1990). These sediments were probably deposited prior to significant uplift of the present-day Oki Ridge. The uplift of the Oki Ridge during late Pliocene/Pleistocene times is indicated by the disappearence of coarse-grained (turbiditic) terrigenous sediments, the decrease in accumulation rates of bulk accumulation rates (Fig. 114), and the increase in carbonate accumulation rates (Fig. 114). The latter started near 1.9 Ma, probably resulting from a gradual emergence of the Oki Ridge above the early Pleistocene calcite compensation depth (Ingle, Suyehiro et al., 1990).

The increased accumulation rates of terrigenous organic carbon in the late Pliocene/Pleistocene partly coincide with increased (eolian) quartz contents (cf. Fig. 123) which may suggest an eolian transport for the organic matter, too. Detailed investigations of accumulation rates of organic and inorganic terrigenous matter will follow to test this hypothesis (deMenocal et al.; Dersch and Stein; Stein and Dunbar; Stein and Stax; all in prep.).

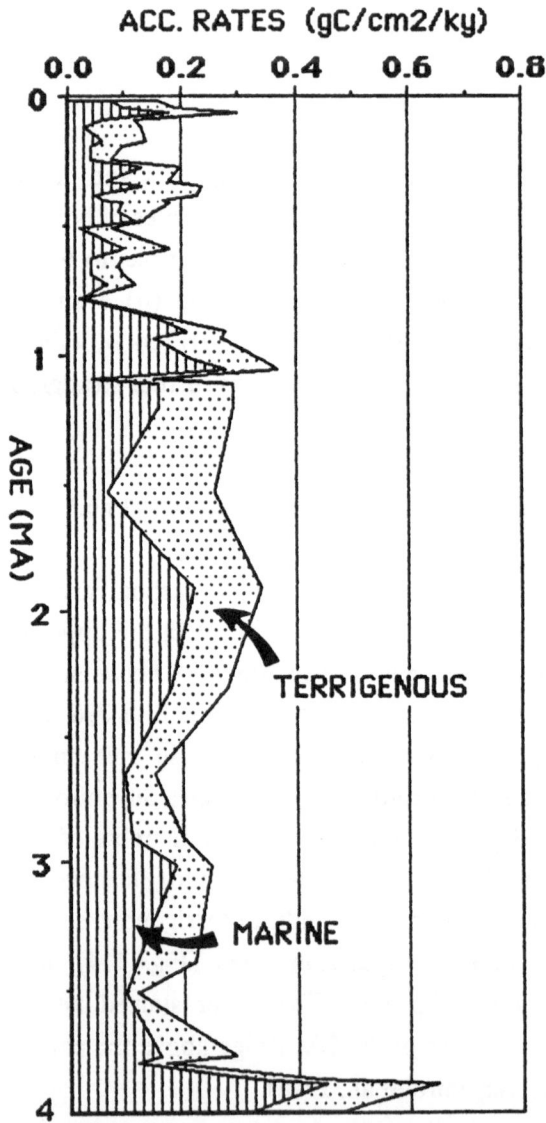

Fig. 121: Accumulation rates of terrigenous and marine organic carbon at Site 798 (preliminary results, based on shipboard stratigraphy and marine/terrigenous organic carbon estimates from Rock-Eval pyrolysis data; cf., Stein et al., 1986).

8.5.2. Changes in accumulation rates of marine organic carbon: High-productivity vs. high-preservation events ?

According to C/N ratios, hydrogen indices, and first maceral data (see above), the organic carbon fraction at Sites 798 and 799 is characterized by high amounts of marine organic material. Accumulation of major amounts of marine organic carbon requires special environmental conditions such as increased surface-water productivity, increased preservation rate of organic matter under anoxic conditions, and/or rapid burial of organic matter by turbidites that also enhanced preservation of organic carbon (cf., Chapter 2, Fig. 3). Since high organic carbon values were recorded throughout the entire sediment sequences of Sites 798 and 799, rapid burial by turbidites which might have occurred ocassionally at Site 798 and certainly more often at Site 799 (Ingle, Suyehiro et al., 1990), cannot be regarded as dominant mechanism controlling the general organic carbon enrichments. Increased preservation in turbidites may only explain single events of organic carbon enrichments. Thus, high surface-water productivity and/or increased preservation rate under anoxic conditions should have mainly caused (marine) organic carbon accumulation at Site 799 and, especially, at Site 798. Furthermore, the long-term changes in organic carbon deposition are certainly influenced by the tectonic evolution of the Japan Sea.

A first idea about these different mechanisms of organic-carbon enrichments may be obtained from a simple carbonate/organic-carbon diagram shown for the middle/late Pleistocene interval of Site 798A (Fig. 122):

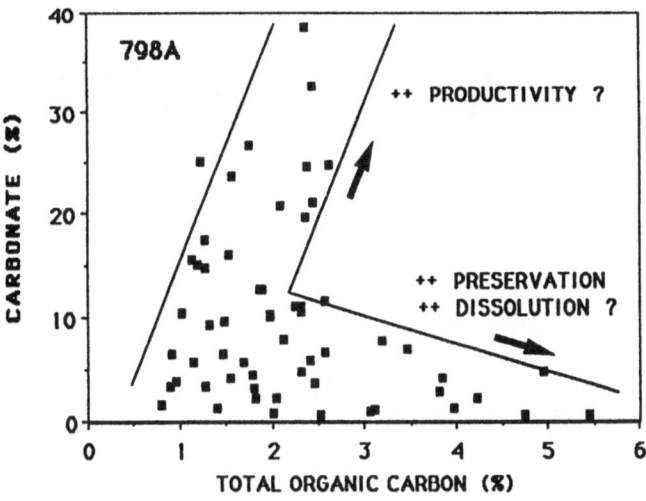

Fig. 122: Correlation between carbonate and organic carbon contents at Site 798A (i.e., the last about 1 Ma).

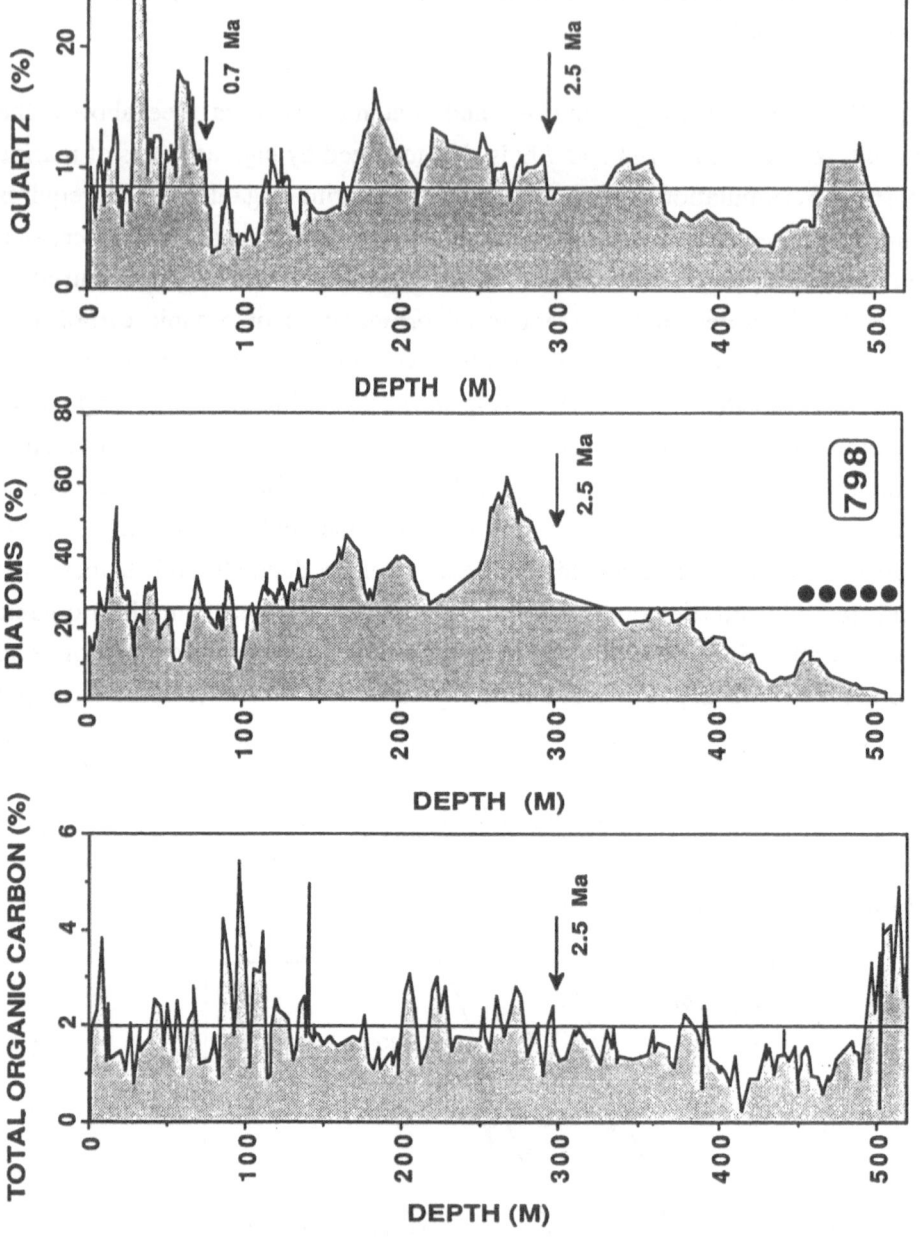

Fig. 123: Total organic carbon, diatom, and quartz contents (%) at Site 798. The diatom and quartz data are based on smear-slide estimates; solid dots indicate occurrence of opal-CT (Ingle, Suyehiro et al., 1990).

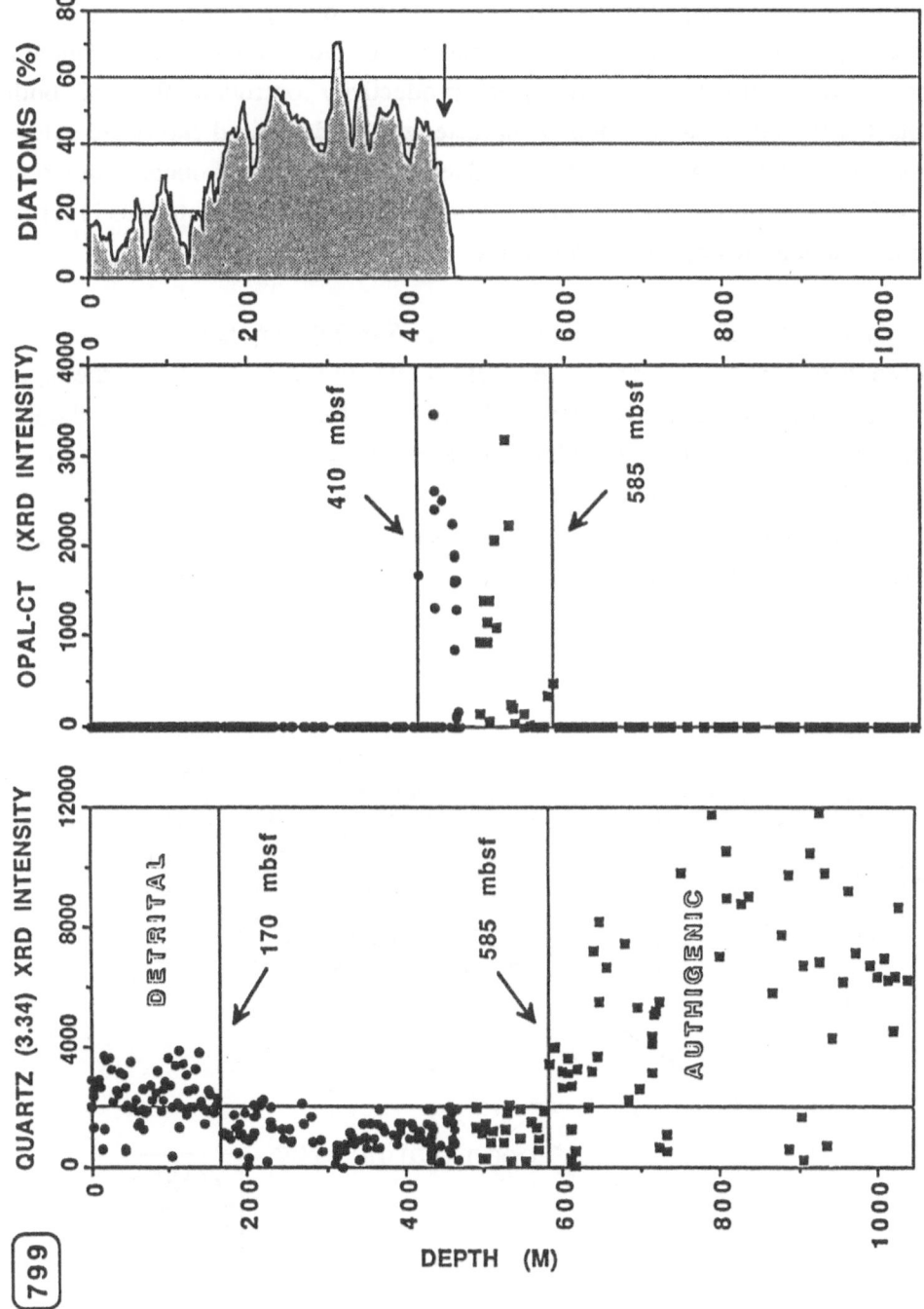

Fig. 124: Occurrence of quartz, opal-CT, and diatoms (opal-A) at **Site 799.** Quartz and opal-CT data are rough data from XRD analyses; diatom data are based on smear-slide estimates (Ingle, Suyehiro et al., 1990).

In this diagram, two groups of data points can be distinguished. Group A is characterized by a parallel increase of both carbonate and organic carbon contents which may have resulted from increased productivity-controlled flux of both components. On the other hand, group B characterized by decreased carbonate values and high organic carbon contents can be explained by increased carbonate dissolution and increased preservation of organic carbon. For more precise interpretations in terms of changes in sediment fluxes, accumulation rates have to be used.

The accumulation rates of marine organic carbon at Site 798 vary between about 0.05 and 0.4 gC cm^{-2} ky^{-1} (Fig. 121) which are values similar to those recorded in the high-productivity upwelling area off Northwest Africa (cf., Figs. 83 and 90). Although accumulation rates have not been calculated for Site 799 because of the occurrence of turbidites and slumps and a still very preliminary stratigraphy (Ingle, Suyehiro et al., 1990), it is definite that average sedimentation rates (and thus also accumulation rates of marine organic carbon) are high at this site, too.

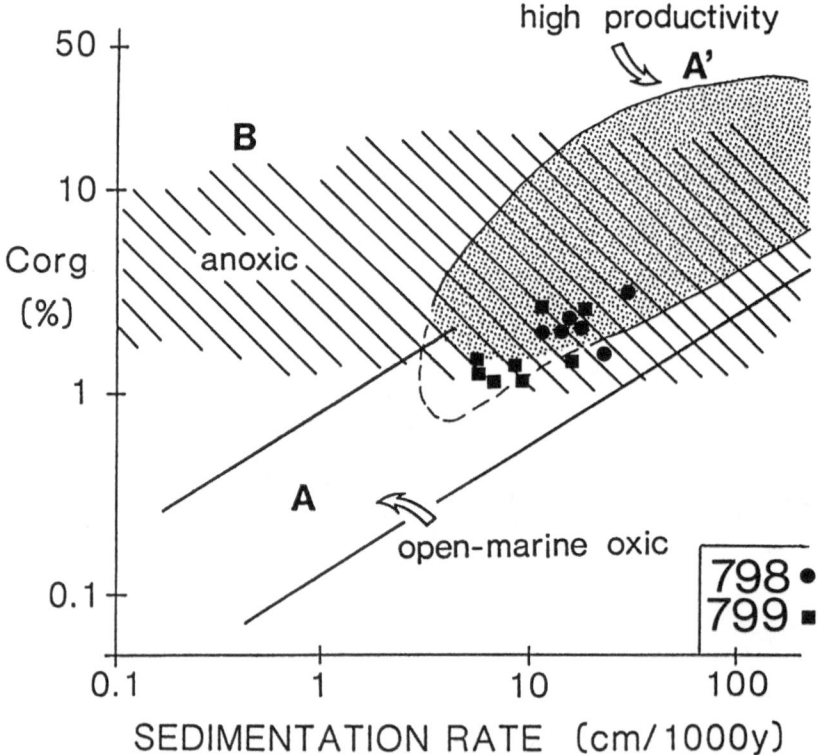

Fig. 125: Correlation between organic carbon content and sedimentation rate at Sites 798 and 799 (according to Stein, 1986b, 1990). For further explanation of fields A, A', and B see Figure 11.

According to smear-slide analyses and XRD measurements, the sediments at Sites 798 and 799 contain high amounts of diatoms and sponge spicules (opal-A), opal-CT, or authigenic quartz (Figs. 123 and 124); the major lithologies are diatomaceous ooze/clay, siliceous claystone, and porcellanite (Ingle, Suyehiro et al., 1990). These types of sediment also indicate a high-surface-water productivity environment (e.g., Isaacs, et al., 1983; Berger et al., 1989; Bohrmann and Stein, 1989). Plotted in an organic-carbon/sedimentation-rate diagram (cf., Fig. 11), the data points of Sites 798 and 799 also fit into the "high-productivity field" (Fig. 125; cf. Figs. 87 and 88 for upwelling areas off Northwest Africa, Peru, and Arabia), although a clear distinction between "high-productivity" and "anoxic environment" is not possible (cf., Stein, 1990).

At Site 798, allmost synchronous increases in (marine) organic-carbon, diatom, and quartz contents at about 300 mbsf (i.e., near 2.5 Ma) suggest an increase in surface-water productivity, paralleled by a climate-controlled increase in (eolian) quartz supply at Oki Ridge during that time (Fig. 123), i.e., during times of development of major Northern Hemisphere Glaciation (e.g., Shackleton et al., 1984) and major expansion of deserts and loess deposition in China (e.g., Burbank and Jijun, 1985; Tungsheng, 1988). The increase of eolian sediment supply from the Asian continent at that time is also recorded in the central North Pacific (Rea and Janecek, 1982). Similar changes, i.e., increased surface-water productivity and increased eolian sediment supply due to intensified atmospheric circulation and aridification also occurred in the eastern subtropical and tropical Atlantic near 2.5 Ma (e.g., Stein, 1985a, 1985b; Ruddiman and Janecek, 1989). Based on smear slide and XRF data of Site 798 sediments, accumulation rates of opal were estimated semi-quantitatively to reach maximum values of 3 to 8 g cm^{-2} ky^{-1} at approximately 2.2 to 2.5 Ma (Ingle, Suyehiro, et al., 1990). These rates are similar to those determined for the modern high-productivity upwelling environment off Peru (Scheidegger and Krissek, 1983). More detailed records on accumulation rates of marine organic carbon, biogenic opal, and (eolian) quartz as well as an oxygen-stable isotope record are required to prove this first interpretation (deMenocal et al.; Dunbar; Stein and Dunbar; Stein and Stax; all in prep.).

Using equation (4), paleoproductivity was found to vary between about 110 and 260 gC m^{-2} y^{-1} in the Japan Sea (Tab. 11). These values are similar to those calculated for sediments from the upwelling area off Northwest Africa (cf., Fig. 89). The Japan Sea values have to be used with caution, however, because equation (4) is based on data derived from open-ocean (sub-) oxic environments (cf., Chapter 3.1.). If anoxic conditions are assumed, distinct (marine) organic carbon enrichments of 4 % can be reached although productivity remained relatively low (Tab. 11: 150 gC m^{-2} y^{-1}, using equation (5); cf., Chapter 3.1.). Based on the relationship between organic-carbon content and (pyritic) sulfur (Leventhal, 1983; Berner, 1984, 1989), these anoxic deep-water conditions should have occurred at least occasionally in the Japan Sea though Miocene to Quaternary times (Figs. 126 and 127).

This interpretation is in agreement with results from DSDP-Leg 31 and piston-core data (Ujiié and Ichikura, 1973; Ingle, 1975; Oba, 1983; Oba et al., 1990). Anoxic conditions should have occurred during times of restricted exchange between the Sea of Japan and the Pacific Ocean (cf., Chapter 8.2. and Fig. 110). Further studies of sediments deposited during "oxic and anoxic periods", including data on the complex sulfur-carbon-iron relationship (cf., Berner, 1989) and oxygen-stable isotopes for stratigraphic correlation, have to be collected, before more precise estimates of paleoproductivity are possible.

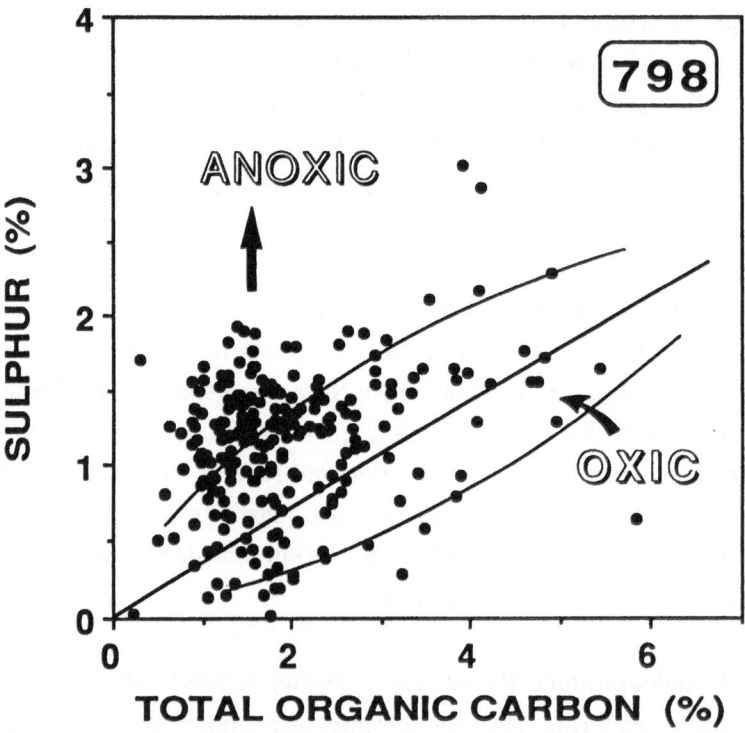

Fig. 126: Relationship between organic carbon and (pyritic) sulfur contents at Site 798. Distinction between oxic and anoxic environment according to Leventhal (1983) and Berner (1984). See also Figure 16.

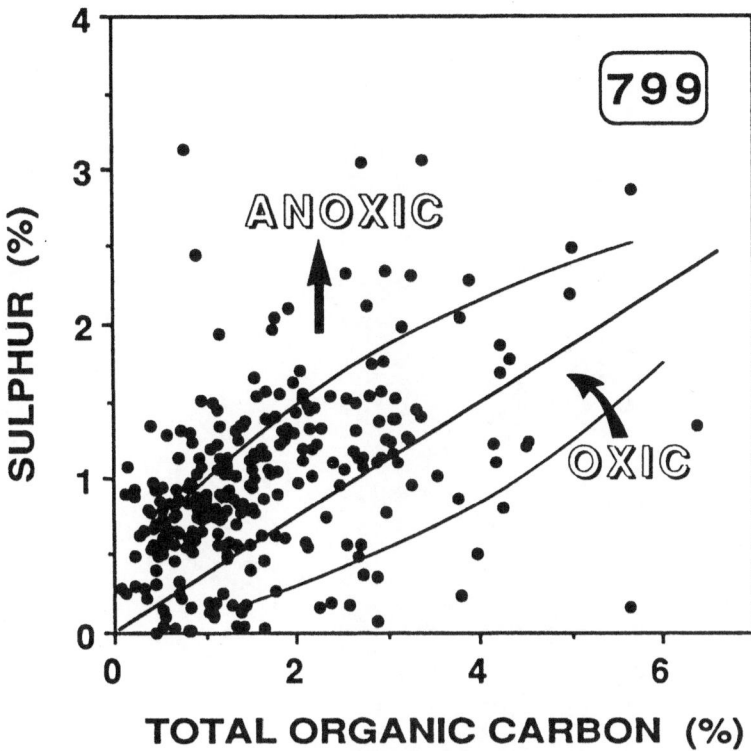

Fig. 127: Relationship between organic carbon and (pyritic) sulfur contents at Site 799. Distinction between oxic and anoxic environments according to Leventhal (1983) and Berner (1984). See also Figure 16.

Tab. 11: Mean paleoproductivity values for Quaternary light bioturbated/homogeneous (1-2 % C_{org}) and dark laminated sediments (3-4 % C_{org}) from Site 798 ; (*) calculated using equation (4) (cf., Stein, 1986a), (+) using equation (5) (cf., Bralower and Thierstein, 1984). Mean sedimentation rate and water depth (900 m) from Ingle, Suyehiro et al. (1990), wet bulk density and porosity data from Holler in Ingle, Suyehiro et al. (1990). See text for further explanation.

Marine C_{org} (%)	LSR (cm/ky)	WBD (g cm^{-3})	PO (%)	Environment	Productivity (gC m^{-2} ky^{-1})
1 - 2	12	1.5	70	oxic	110 - 180 (*)
3 - 4	12	1.4	74	(sub-)oxic	215 - 260 (*)
3 - 4	12	1.4	74	anoxic	120 - 150 (+)

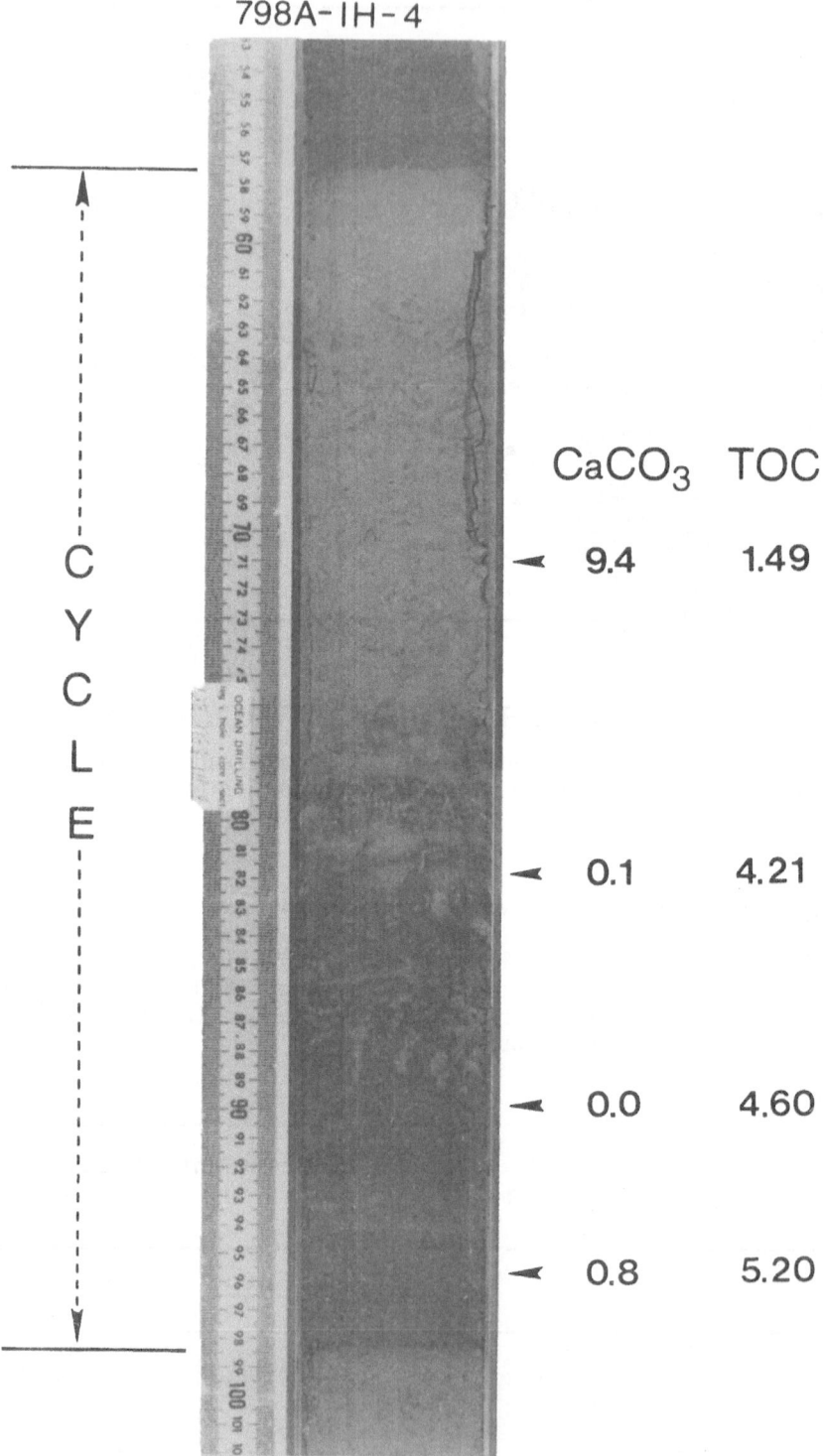

Fig. 128: Dark/light cycle at Site 798A (798A-1H-4, 55-100 cm). On the right-hand side, carbonate and total organic carbon contents (%) are shown.

A key to answer the question about mechanisms causing the organic carbon enrichments at Site 798 (i.e., high-productivity vs. high-preservation events), is certainly a detailed investigation of the remarkable dark/light cycles preserved in the Quaternary sediments of this site (Fig. 127; Ingle, Suyehiro et al., 1990; Föllmi et al., 1990). In general, the dark intervals are finely laminated and characterized by increased organic carbon contents of up to 6 %, whereas the lighter intervals are homogeneous or bioturbated and display organic carbon contents of about 2 % (Fig. 127). The thickness of these cycles vary between 10 and 100 cm which corresponds to periods of about 1000 to 12000 years (based on mean sedimentation rates; cf., Fig. 111). This means, these cycles have periods distinctly lower than those of the Milankovitch-type cycles.

Tab. 12: Carbonate, organic carbon, total nitrogen, total sulfur, and quartz contents (wt% of bulk sediment), C/N ratio, hydrogen index (HI) values (mgHC/gC), oxygen index (OI) values (mgCO$_2$/gC), grain-size distribution (%), quartz/feldspar (Q/F) ratio, and composition of clay fraction (% of clay minerals) (Ill = illite; Chl = chlorite; Kao = kaolinite; Smec = smectite) of the dark/light cycle in Section 798B-10-7. Samples No. 1 and 6 are from light sediments, No. 2 to 5 form dark sediments.

Sample	Color	CaCO$_3$	C$_{org}$	N$_{tot}$	S$_{tot}$	C/N	HI	OI
1. 10-7-025	light	4.2	2.32	0.24	1.21	10	196	62
2. 10-7-054	dark	0.7	4.67	0.36	1.55	13	403	54
3. 10-7-074	dark	8.1	4.61	0.36	1.76	13	333	42
4. 10-7-075	dark	12.7	4.08	0.32	1.29	13	349	59
5. 10-7-085	dark	2.7	4.82	0.40	1.71	12	399	46
6. 10-7-091	light	2.7	2.94	0.28	1.62	11	403	57

Sample	Sand	Silt	Clay	Q	Q/F	Ill	Chl	Kao	Smec
10-7-025	1	96	3	27	2.4	75	18	7	tr
10-7-054	0	88	12	24	2.2	67	29	4	tr
10-7-074	0	99	1	10	1.9	71	29	tr	tr
10-7-075	0	83	17	9	1.4	66	34	tr	tr
10-7-085	8	77	15	20	2.2	65	35	tr	tr
10-7-091	1	93	2	70	2.1	69	26	5	tr

For a first cycle (128-798B-10-7), some detailed data are already available, including carbonate, organic carbon, and biogenic opal contents, hydrogen and oxygen indices as well as XRD data (Figs. 129 and 130; Tab. 12). Based on these results, a preliminary interpretation of the organic carbon enrichment is possible:

First, the Rock-Eval data clearly indicate the marine origin of the organic material throughout the cycle (Fig. 129). In the lower part of the cycle, the dark laminated sediments are characterized by increasing contents of carbonate and biogenic opal, and a high marine organic carbon content (Fig. 130), which may suggest increased productivity and increased organic carbon flux causing anoxic conditions. These anoxic conditions may have already developed before times of maximum productivity. In the middle part, carbonate content and (although less distinct) biogenic opal content decrease whereas the marine organic carbon content remains high. This may reflect decreasing productivity, but still high preservation rate of organic carbon under anoxic deep-water conditions.

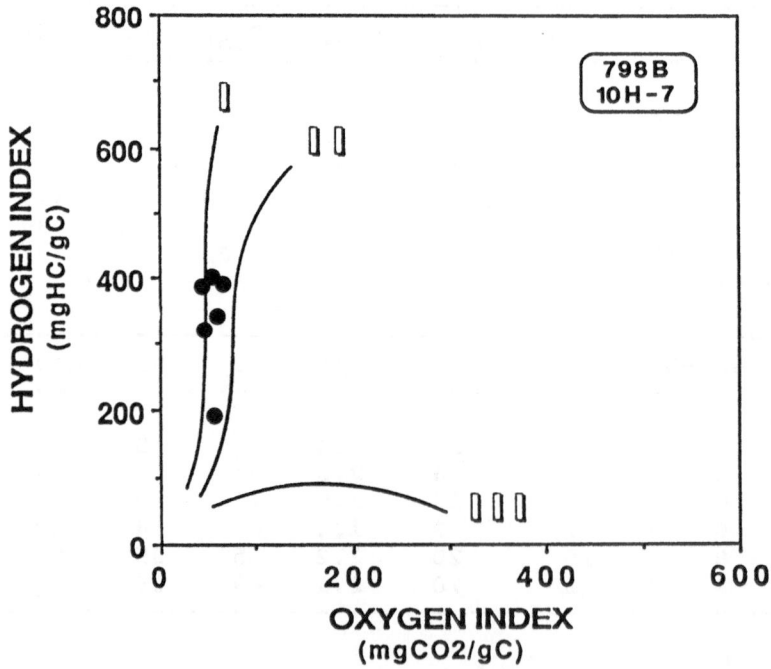

Fig. 129: Hydrogen and oxygen index values of sediments from a dark/light cycle at Site 798B (Core 10H-7; cf. Tab. 11).

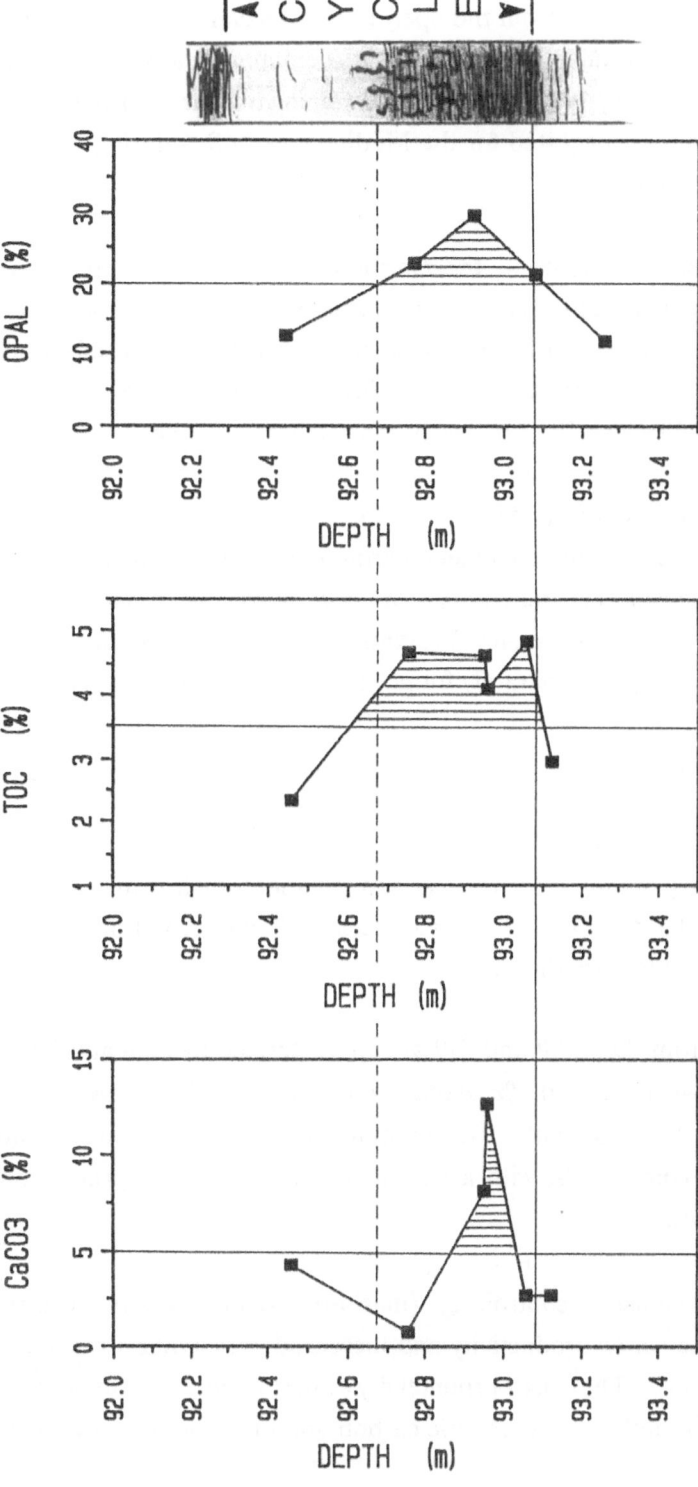

Fig. 130: Carbonate, total organic carbon, and biogenic opal in the dark/light cycle of 798B-10-7 (cf., Tab. 11). Biogenic opal data from Dunbar (unpubl.).

These anoxic conditions may have occurred during times of lowered sea level when the Japan Sea was more restricted from the open Pacific Ocean and the oceanic circulation and the ventilation of the deep-water sphere in the Japan Sea were distinctly reduced (cf., Oba, 1983; Oba et al., 1990). The deep-water environment in the Japan Sea may have been similar to that suggested for the Mediterranean Sea during times of sapropel formation (cf., Chapter 7.).

In the upper light part of the cycle, the marine organic carbon content and hydrogen indices also decrease (Fig. 130, Tab. 12), indicating the development of more oxidizing conditions in the Japan Sea during times of higher sea level, re-establishment of the connection to the Pacific Ocean, and intensification of oceanic circulation (cf., Oba, 1983; Oba et al., 1990). The light intervals are furthermore characterized by a higher quartz content and a slightly increased content of kaolinite (Tab. 12) which point to a different source area and/or different climatic conditions in the source area of the terrigenous material. For detailed paleoenvironmental reconstructions, oxygen-stable isotope data are absolutely necessary for correlation with global climate, i.e., glacial/interglacial or low/high sea-level conditions, as well as accumulation rates of the different variables calculated separately for glacial and interglacial intervals.

8.6. Conclusions

Results of preliminary organic-geochemical and sedimentological investigations yielded informations about the depositional and paleoceanographic evolution in the Japan Sea during Miocene to Quaternary times.

(1) The sediments from Sites 798 and 799 are characterized by high amounts of organic carbon, ranging between 0.5 and 6%. Accumulation rates of up 0.4 gC cm^{-2} ky^{-1} were calculated for Site 798 sediments. The organic matter is a mixture of marine and terrigenous organic compounds, with a dominance of the marine proportion in major parts of the sequences.

(2) Dominant mechanisms controlling (marine) organic-carbon enrichments are probably high-surface-water productivity and increased preservation rates under anoxic deep-water conditions. The high-productivity environment is indicated by high accumulation rates of both marine organic carbon and biogenic opal. Based on organic-

carbon/sulfur relationships, anoxic conditions should have occasionally occurred in the Japan Sea.

(3) In the uppermost lower Pliocene sediments at Site 798 and the Quaternary to Miocene sediments at Site 799, rapid burial of organic carbon in turbidites may have occurred episodically.

(4) Remarkble cycles of dark laminated sediments with maximum organic carbon content of more than 5 % and light bioturbated or homogenoeus sediments with lower organic carbon content indicate dramatic short-term paleoceanographic variations, best developed during middle/late Pleistocene times. First comparisons of carbonate, opal, and organic carbon data suggest a succession of "anoxic and high-productivity conditions - anoxic and reduced-productivity conditions - oxic conditions" causing some of the cycles at Site 798.

(5) More detailed records on accumulation rates of marine and terrigenous organic carbon, carbonate, and biogenic opal and the correlation with oxygen-stable isotopes (i.e., the correlation with glacial/interglacial and low/high-sealevel cycles) are absolutely necessary for a more precise and complete reconstruction of the environmental history of the Japan Sea.

9. Summary: Organic carbon accumulation in late Cenozoic sediments from ODP-Legs 105, 108, and 128.

According to the results of detailed organic-geochemical and sedimentological investigations of sediments drilled during ODP-Legs 105, 108, and 128, sediments characterized by high organic carbon contents were deposited in very different marine environments. The accumulation rates of total organic carbon are also in the same order of magnitude and vary between 0.05 and 0.4 gC cm^{-2} ky^{-1} at Site 658 (upwelling area off Northwest Africa), 0.1 and 0.5 gC cm^{-2} ky^{-1} at Site 798 (Japan Sea), and 0.04 and 0.2 gC cm^{-2} ky^{-1} at Site 645 (Baffin Bay). From this it is obvious that accumulation rates of total organic carbon (i) do not allow to distinguish between different factors controlling organic carbon enrichment (cf., Fig. 3) and (ii) cannot be used for detailed reconstruction of the depositional environment without significant information about the composition of the organic carbon fraction. On the other hand, data on both quantity and quality of the organic matter give important insights into the depositional history of the sediments and allow reconstructions of the evolution of paleoclimate, paleoceanic circulation, and surface-water paleoproductivity (Fig. 131). Furthermore, estimates of the source rock potential for gas and/or oil are possible, too.

- In Baffin Bay, the supply of terrigenous organic carbon was dominant. The changes reflect distinct fluctuations of the climate in the surrounding continents.

- In the upwelling area off Northwest Africa (and also off Peru and Arabia) high productivity was the major factor controlling (marine-) organic matter deposition (Fig. 132A) whereas the supply of terrigenous organic matter was of secondary importance. Occasionally, the organic-carbon-rich sediments were redeposited and preserved as turbidites in the deep-sea environment.

- In the Japan Sea, both high productivity (Fig. 132A) and high preservation rate under anoxic conditions (Fig. 132B) may have resulted in (marine-) organic-carbon-rich sediments. High terrigenous organic carbon supply and rapid burial by turbidites may have occurred episodically.

In all of these environments, organic-carbon enrichment is clearly triggered by long-term as well as short-term changes of the global climate (cf., conclusions of Chapters 5 to 8). However, there is no simple relationship between climate and organic carbon supply (or productivity). This means, it is not possible to postulate that productivity was generally higher during glacial times and lower during interglacial times or vice versa (cf.,Sarnthein et al., 1987; Mix, 1989; Stein and Littke, 1990). Instead, local phenomena such as sea-ice cover (cf., Chapter 5.5.3.) and fluvial nutrient supply (cf., Chapter 6.5.2.) which varied with time and area, may have influenced surface-water productivity significantly. For example, changes in the relative importance between upwelling activity (which was increased during glacials) and fluvial nutrient supply off Northwest Africa (which was increased during interglacials) may have been different for different glacial/interglacial cycles, resulting in the complex productivity record at Site 658. These results make calculations of changes in global paleoproductivity (and global CO_2 budgets) more complicated.

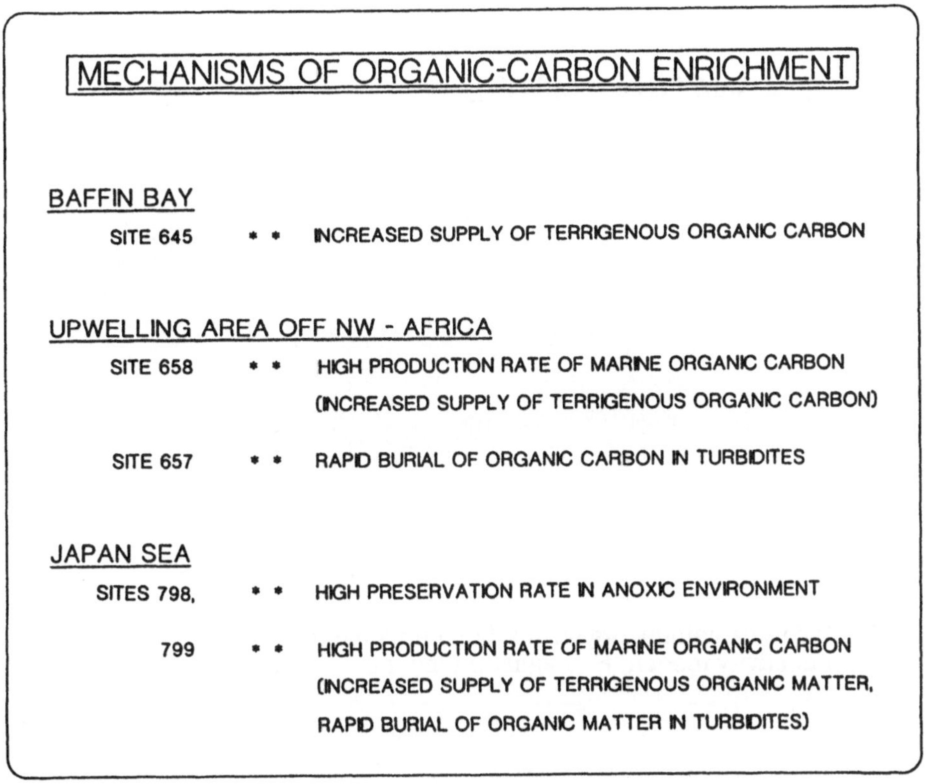

MECHANISMS OF ORGANIC-CARBON ENRICHMENT

BAFFIN BAY
    SITE 645   * *   INCREASED SUPPLY OF TERRIGENOUS ORGANIC CARBON

UPWELLING AREA OFF NW - AFRICA
    SITE 658   * *   HIGH PRODUCTION RATE OF MARINE ORGANIC CARBON
                    (INCREASED SUPPLY OF TERRIGENOUS ORGANIC CARBON)

    SITE 657   * *   RAPID BURIAL OF ORGANIC CARBON IN TURBIDITES

JAPAN SEA
    SITES 798,   * *   HIGH PRESERVATION RATE IN ANOXIC ENVIRONMENT

        799   * *   HIGH PRODUCTION RATE OF MARINE ORGANIC CARBON
                    (INCREASED SUPPLY OF TERRIGENOUS ORGANIC MATTER,
                    RAPID BURIAL OF ORGANIC MATTER IN TURBIDITES)

Fig. 131: Mechanisms of organic carbon enrichments dominating in Baffin Bay, the upwelling area off northwest Africa, and the Japan Sea.

A. PRODUCTIVITY MODEL

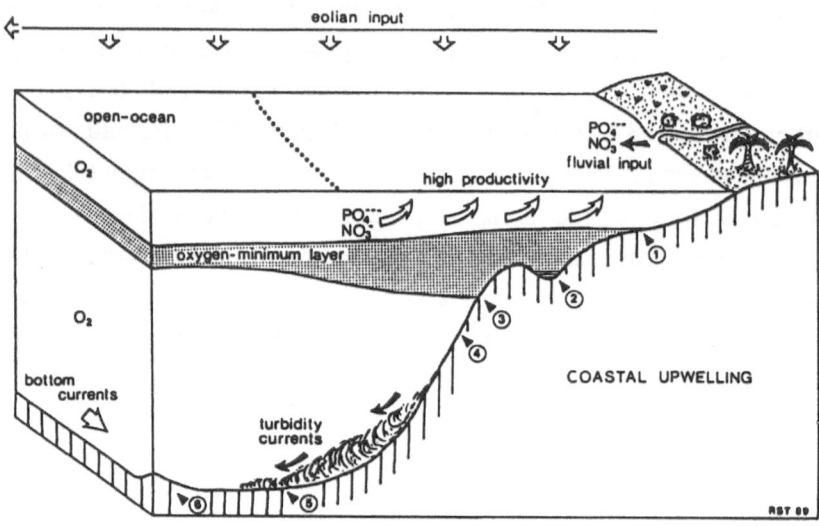

B. STAGNATION MODEL

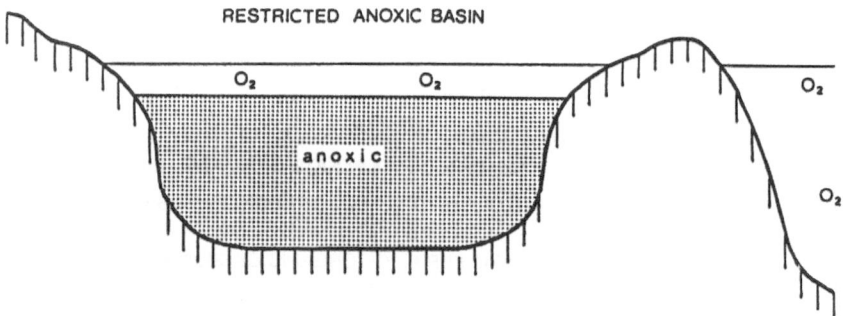

Fig. 132: Models explaining organic carbon enrichments.
A. High-productivity model; B. Stagnation model.

10. Epilogue: Organic carbon accumulation in the Cretaceous Atlantic

The results of the three case studies (Leg 105, 108, and 128) on mechanisms controlling organic carbon enrichment may also help to reconstruct the formation of fossil carbonaceous sedimentary rocks, such as the Cretaceous black shales. During that time, a widespread (global) occurrence of black shales in marginal as well as open-ocean environments was recorded. Although the global paleogeography, paleoclimate, paleoceanic circulation, and paleocean chemistry were different in the mid-Cretaceous in comparison to the Present, (Schlanger and Jenkyns, 1976; Herbin et al., 1986; Stein et al., 1986; Arthur et al., 1988), the factors described in Figure 3 may explain differences in quantity and quality of organic carbon recorded in black shales from different oceanic basins and different Cretaceous stages (Fig. 133; cf., Stein et al., 1986, 1989c):

Hauterivian black shales from the western North Atlantic characterized by the dominance of terrigenous organic matter (Fig. 134A; for accumulation rates see Stein et al., 1986), probably were caused by increased supply of terrigenous organic carbon (i.e., a situation similar to that of Baffin Bay).

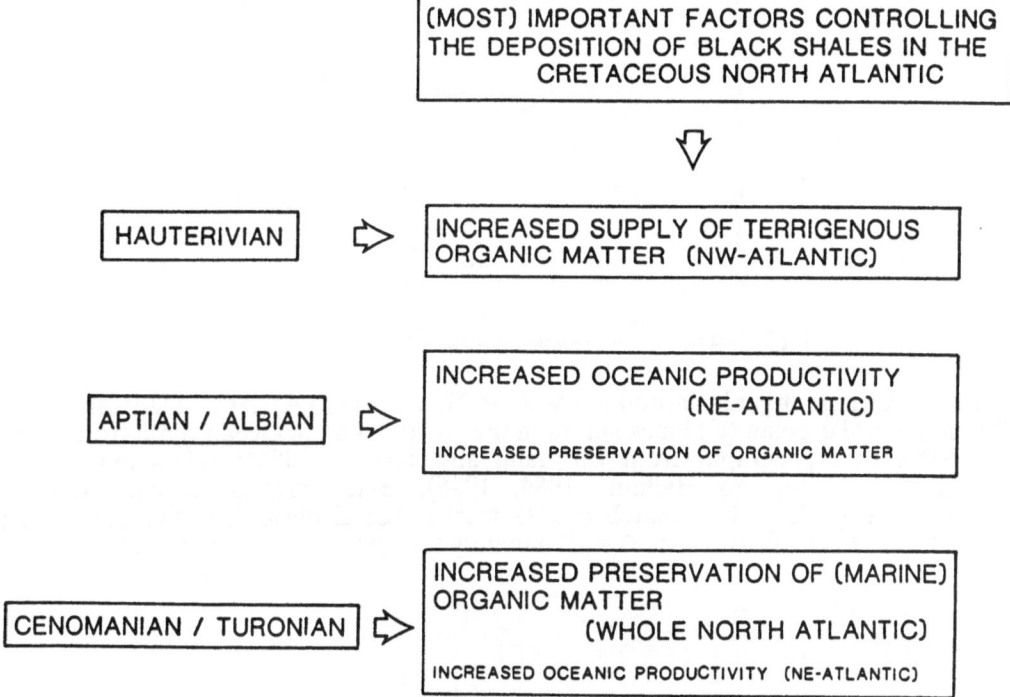

Fig. 133: Mechanisms controlling organic carbon accumulation in the Cretaceous Atlantic.

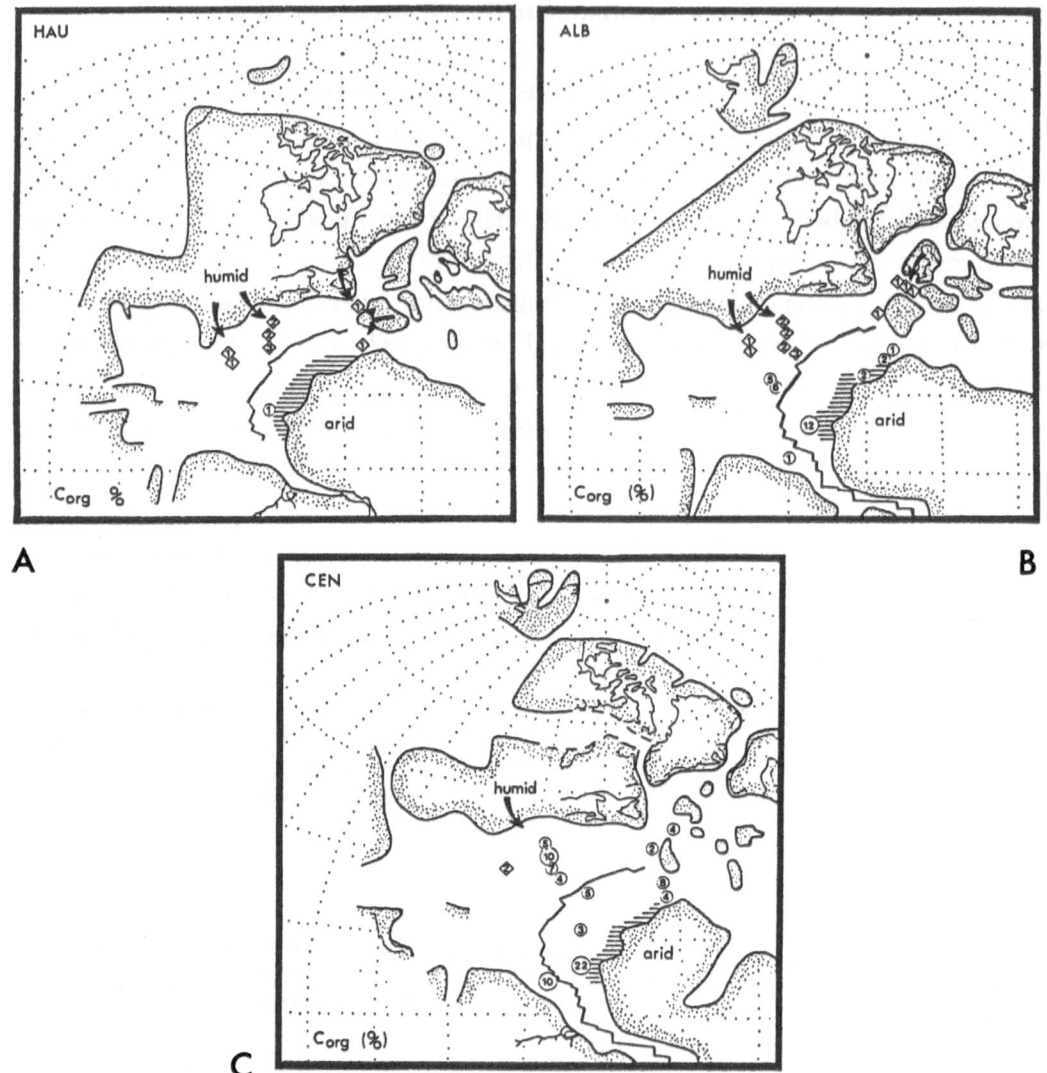

Fig. 134: Average organic carbon contents in Hauterivian (A), Aptian/Albian (B), and Cenomanian/Turonian (C) black shales in the North Atlantic Ocean (data from Stein et al., 1986). Paleogeography from Ehrmann and Thiede (1985); paleoclimate on the continents according to Hallam (1984, 1986). Black arrows indicate supply of terrigenous organic matter. Hatched area marks coastal upwelling. Numbers indicate percentages of organic carbon; ◇ = dominantly terrigenous, ○ = dominantly marine.

On the other hand, Aptian/Albian black shales from the South Atlantic and the eastern North Atlantic and Cenomanian/Turonien black shales from the entire North Atlantic characterized by high amounts of marine organic carbon (Fig. 134B and C; Stein et al., 1986), probably resulted from increased surface-water productivity and/or anoxic deep-water conditions (i.e., situations similar to those described for the upwelling area off Northwest Africa and the Japan Sea). The latter (i.e., ocean-wide to global "stagnation") may have been the most important mechanism controlling formation of Cenomanian/Turonian black shale (e.g., Brumsack, 1980; Bralower and Thierstein, 1984; de Graciansky et al., 1984; Herbin et al., 1986; Stein, 1986b). It has to be mentioned, however, that the depositional history of these organic-carbon-rich strata still is a matter of controversy (i.e., "high-productivity" vs. "deep-water stagnation" models, Fig. 132; e.g., Thiede and van Andel, 1977; Herbin et al., 1986; Stein et al., 1986; Arthur et al., 1988) and further multidisciplinary studies are necessary to solve the black shale problem (Fig. 135).

FUTURE WORK

TO DECIPHER GLOBAL AND REGIONAL/LOCAL
PALEOCEANOGRAPHIC AND PALEOCLIMATIC CONDITIONS
CAUSING THE FORMATION OF MESOZOIC BLACK SHALES
WE NEED:

••• MULTIDISCIPLINARY STUDIES
(ORGANIC/INORGANIC GEOCHEMISTRY, SEDIMENTOLOGY, PALEONTOLOGY)

••• HIGH-RESOLUTION BIOSTRATIGRAPHIC FRAMEWORK

••• SAMPLES FROM DIFFERENT OCEANIC ENVIRONMENTS
(DEEP SEA TO EPICONTINENTAL SEAS)
AND FROM ALL OCEANS
(ESPECIALLY THE SUPER-PACIFIC AND ARCTIC)

••• DETAILED COMPARISON OF APTIAN/ALBIAN AND
CENOMANIAN/TURONIAN INTERVALS

Fig. 135: Future work on black shales.

11. References

Agwu, C.O.C. and Beug, H.J., 1982. Palynological studies of marine sediments off the West African coast, 'Meteor'-Forsch. Ergebn., C, Vol. 36, p. 1-30.

Aksu, A.E., 1981. Late Quaternary stratigraphy, paleoenvironmentology, and sedimentation history of Baffin Bay and Davis Strait. Ph.D. dissertation, Dalhousie University, Halifax, Nova Scotia, Canada, 585 pp.

Aksu, A.E. and Hillaire-Marcel, C., 1989. Upper Miocene to Holocene oxygen and carbon isotopic stratigraphy of Sites 646 and 647, Labrador Sea. In: Srivastava, S.P., Arthur, M.A., et al., Proc. ODP, Sci. Results, 105, College Station, Tx (Ocean Drilling Program), p. 689-704.

Aksu, A.E. and Kaminski, M.A. 1989. Neogene and Quaternary planktonic foraminifer biostratigraphy and biochronology in Baffin Bay and the Labrador Sea. In: Srivastava, S.P., Arthur, M.A., et al., Proc. ODP, Sci. Results, 105; College Station, Tx (Ocean Drilling Program), p. 287-304.

Aksu, A.E., de Vernal, A., and Mudie, P.E., 1989. High-resolution foraminifer, palynologic, and stable isotopic records of upper Pleistocene sediments from the Labrador Sea: Paleoclimatic and paleoceanographic trends. In: Srivastava, S.P., Arthur, M.A., et al., Proc. ODP Sci. Results, 105, College Station, Tx (Ocean Drilling Program), p. 617-652.

Aller, R.C. and Mackin, J.E., 1984. Preservation of reactive organic matter in marine sediments. Earth Pla. Sci. Lett., Vol. 70, p. 260-266.

Altenbach, A.V., 1985. Die Biomasse der benthischen Foraminiferen. Auswertungen von 'Meteor'-Expeditionen im östlichen Nordatlantik. Ph.D. dissertation, University of Kiel.

Altenbach, A.V. and Sarnthein, M., 1989. Productivity Record in Benthic Foraminifera. In: Berger, W.H., et al. (Eds.), Productivity of the Ocean: Past and Present, Life Sci. Res. Rep., Vol. 44, Wiley & Sons, p. 255-270.

Arthur, M.A., Dean, W.E., and Claypool, G.E., 1985. Anomalous ^{13}C enrichment in modern marine organic carbon. Nature, Vol. 315, p. 216-218.

Arthur, M.A., Dean, W.E., and Stow, D.A.V., 1984. Models for the deposition of Mesozoic - Cenozoic fine-grained organic-carbon-rich sediments in the deep sea. In: Stow, D.A.V. and Piper, D.J.W. (Eds.), Fine-Grained Sediments: Deep-Water Processes and Facies. Geol. Soc. London, Spec. Publ., Vol. 15, p. 527-560.

Arthur, M.A., Jenkyns, H., Brumsack, H., and Schlanger, S., 1988. Stratigraphy, Geochemistry, and Paleoceanography of organic-carbon-rich mid-Cretaceous sequences. In: Beaudoin, B. and Ginsburg, R. (Eds.), Cretaceous Resources, Events, and Rhythms, Digne, France, p. 25-70.

Arthur, M.A., Srivastava, S.P., Kaminski, M.A., Jarrad, R., and Osler, J., 1989. Seismic stratigraphy and history of deep circulation and sediment drift development in Baffin Bay and the Labrador Sea. In: Srivastava, S.P., Arthur, M.A., et al., Proc. ODP, Sci. Results, 105, College Station, Tx (Ocean Drilling Program), 957-988.

Bainbridge, A.E., 1976. GEOSECS Atlantic Expedition, Vol. 2, Sections and Profiles.

Baldauf, J.G., Clement, B.G., Aksu, A.E., de Vernal, A., Firth, J.V., Hall, F., Head, M.J., Jarrad, R.D., Kaminski, M.A., Lazarus, D., Monjanel, A.L., Berggren, W.A., Gradstein, F.E., Knüttel, S., Mudie, P.J., and Russell, M.D., 1989. Magnetostratigraphic and biostratigraphic synthesis of Ocean Drilling Program Leg 105: Labrador Sea and Baffin Bay. In: Srivastava, S.P., Arthur, M.A., et al., Proc. ODP, Sci. Results, 105, College Station, Tx (Ocean Drilling Program), p. 935-956.

Berger, W.H., Fischer, K., Lai, C., and Wu, G., 1987. Ocean productivity and organic carbon flux. I. Overview and maps of primary production and export production. Univ. California, San Diego, SIO Reference 87-30.

Berger, W.H. and Keir, R.S., 1984. Glacial-Holocene changes in atmospheric CO_2 and the deep-sea record. In: Hansen, J.H. and Takahashi, T. (Eds.), Climate Processes and Climate Sensitivity, Geophys. Monog., Vol. 29, p. 337-351.

Berger, W.H. and Killingley, J., 1977. Glacial-Holocene transition in deep-sea carbonates: selective dissolution and the stable isotope signal. Science, Vol. 197, p. 563-566.

Berger, W.H., Smetacek, V., and Wefer, G., 1989. Productivity of the Ocean: Past and Present. Life Sciences Research Report, Vol. 44, Wiley & Sons, New York, 471 p.

Berner, R.A., 1984. Sedimentary pyrite formation: An update. Geochim. Cosmochim. Acta, Vol. 48, p. 605-615.

Berner, R.A., 1989. Biogeochemical cycles of carbon and sulfur and their effect on atmospheric oxygen over Phanerozoic time. Palaeogeogr., Palaeoclim. Palaeoecol., Vol. 75, p. 97-122.

Betzer, P.R., Showers, W.J., Laws, E.A., Winn, C.D., Ditullo, G.R., and Kroopnick, P.M., 1984. Primary productivity and particle fluxes on a transect of the equator at $153^{\circ}W$ in the Pacific Ocean. Deep-Sea Res., Vol. 31, p. 1-11.

Bishop, J.K.B., 1988. The barite-opal-organic carbon association in oceanic particulate matter. Nature, Vol. 233, p. 241-243.

Bohrmann, G., 1988. Zur Sedimentationsgeschichte von biogenem Opal im nördlichen Nordatlantik und dem Europäischen Nordmeer (DSDP/ODP-Bohrungen 408, 642, 643, 644, 646 und 647). Ber. Sonderforschungsbereich 313, Universität Kiel, Vol. 9, 211 pp.

Bohrmann, G. and Stein, R., 1989. Biogenic silica at ODP-Site 647 in the Southern Labrador Sea: Occurrence, diagenesis, and paleoceanographic implications. In: Srivastava, S.P., Arthur, M.A., et al., Proc. ODP, Sci. Results, 105, College Station, Tx (Ocean Drilling Program), 155-170.

Bohrmann, G., Henrich, R., and Thiede, J., 1990. Miocene to Quaternary Paleoceanography in the Northern Atlantic: Variability in Carbonate and Biogenic Opal Accumulation. In: Bleil, U. and Thiede, J. (Eds.), Geological History of the Polar Oceans: Arctic versus Antarctic. Nato ASI Series C, Vol. 308, Kluwer Acad. Publ., p. 647-675.

Bordowskiy, O.K., 1965a. Sources of organic matter in marine basins. Mar. Geol., Vol. 3, p. 5-31.

Borwoskiy, O.K., 1965b. Accumulation of organic matter in bottom sediments. Mar. Geol., Vol. 3, p. 33-82.

Boyle, E.A., 1986. Deep ocean circulation, preformed nutrients, and atmospheric carbon dioxide: theories and evidence from oceanic sediments. In: Hsü, K.J. (Ed.), Mesozoic and Cenozoic Oceans, AGU Geodynam. Ser., Vol. 15, p. 49-59.

Bralower, T.J. and Thierstein, H.R., 1984. Low productivity and slow deep-water circulation in Mid-Cretaceous oceans. Geology, Vol. 12, p. 614-618.

Brassell, S.C. and Eglinton, G., 1983. The potential of organic geochemical compounds as sedimentary indicators of upwelling. In: Suess, E. and Thiede, J. (Eds.), Coastal Upwelling, Its Sediment Record, Part A, Plenum Press, New York, p. 545-572.

Brassell, S.C., Eglinton, G., Marlowe, I.T., Pflaumann. U., and Sarnthein, M., 1986. Molecular stratigraphy: a new tool for climatic assessment. Nature, Vol. 320, p. 129-133.

Brumsack, H.J., 1980. Geochemistry of Cretaceous black shales from Alantic Ocean (DSDP-Legs 11, 14, 36, and 41). Chem. Geol., Vol. 31, p. 1-25.

Brumsack, H.J., 1986. The inorganic geochemistry of Cretaceous black shales (DSDP-Leg 41) in comparison to modern upwelling sediments from the Gulf of California. In: Summerhayes, C.P. and Shackleton, N.J. (Eds.), North Atlantic Paleoceanography, Geol. Soc. Spec. Publ., Vol. 21, London, p. 447-462.

Bruland, K.W., Bienfang, P.K., Bishop, J.K.B., Eglinton, G., Ittekot, V.A.W., Lampitt, R., Sarnthein, M., Thiede, J., Walsh, J.J., and Wefer, G., 1989. Flux to the Seafloor. In: Berger, W.H., et al. (Eds.), Productivity of the Ocean: Past and Present, Life Sci. Res. Rep., Vol. 44, Wiley & Sons, p. 193-216.

Buggisch, W., 1972. Zur Geologie und Geochemie der Kellwasserkalke und ihrer begleitenden Sedimente (unteres Oberdevon). Abh. Hess. L.-Amt Bodenforsch., Vol. 62, 68 pp.

Burbank, D.W., and Jijun, L., 1985. Age and paleoclimatic significance of the loess of Lanzhou, north China. Nature, Vol. 316, p. 429-431.

Burckle, L.H. and Akiba, F., 1977. Implication of late Neogene freshwater sediment in the Sea of Japan. Geology, Vol. 6, p. 123-127.

Calvert, S.E., 1983. Geochemistry of Pleistocene sapropels and associated sediments from the Eastern Mediterranean. Oceanologica Acta, Vol. 6, p. 255-267.

Calvert, S.E. and Price, N.B., 1983. Geochemistry of Namibian shelf sediments. In: Suess, E. and Thiede, J. (Eds.), Coastal Upwelling, Its Sediment Record, Part A, Plenum Press, New York, p. 337-376.

Calvert, S.E., Vogel, J.S., and Southon, J.R., 1987. Carbon accumulation rates and the origin of the Holocene sapropel in the Black Sea. Geology, Vol. 15, p. 918-921.

Canfield, D.E., 1989. Sulfate reduction and oxic respiration in marine sediments: implications for organic carbon preservation in euxinic environments. Deep-Sea Res., Vol. 36, p. 121-138.

Carlson, T.N. and Prospero, J.M., 1977. Saharan air outbreaks: Meteorology, aerosols, and radiation. Rep. US Gate Centr. Prog. Workshop (NCAR), Boulder, Colorado, p. 57-78.

Chamley, H., 1989. Clay Sedimentology. Springer Verlag Berlin, 623 pp.

Chough, S.K. and Hesse, R., 1987. The Northwest Atlantic Mid-Ocean Channel of the Labrador Sea. 4. Petrography and provenance of the sediments. Can. Journ. Earth Sci., Vol. 24, p. 731-740.

CLIMAP, 1976. The surface of the Ice-Age Earth. Science, Vol. 191, p. 1131-1144.

CLIMAP, 1981. Seasonal reconstructions of the earth's surface at the last glacial maximum. Geol. Soc. Amer., Map and Chart Ser., MC-36.

Colley, S., Thompson, J., Wilson, T.R.S., and Higgs, N.C., 1984. Post-depositional migration of elements during diagenesis in brown clay and turbidite sequences in the Northeast Atlantic. Geochim. Cosmochim. Acta, Vol. 48, p. 1223-1235.

Cornford, C., Rullkötter, J., and Welte, D.H., 1979. Organic geochemistry of DSDP-Leg 47a, Site 397, eastern North Atlantic: Organic petrography and extractable hydrocarbons. In: von Rad, U., Ryan, W.B.F., et al., Init. Reps. DSDP, Vol. 47, Pt. 1, p. 511-522.

Curry, W.B., Duplessy, J.C., Labeyrie, L., and Shackleton, N.J., 1988. Changes in the distribution of δ^{13}C of deep water CO_2 between the last glaciation and the Holocene. Paleoceanography, Vol. 3, p. 317-341.

Curry, W.B. and Lohman, G.P., 1983. Reduced advection into Atlantic Ocean deep eastern basins during last glaciation maximum. Nature, Vol. 306, p. 577-580.

Curry, W.B. and Lohman, G.P., 1985. Carbon deposition rates and deep water residence time in the equatorial Atlantic Ocean throughout the last 160,000 years. In: Sundquist, E.T. and Broecker, W.S. (Eds.), The Carbon Cycle and Atmospheric CO_2: Natural Variations Archean to Present, AGU, Washington, p. 285-303.

Curry, W.B. and Miller, K.G., 1989. Oxygen and carbon isotopic variations in Pliocene benthic foraminifers of the Equatorial Atlantic. In: Ruddiman, W.F., Sarnthein, M., et al., Proc. ODP, Sci. Results, 108, College Station, Tx (Ocean Drilling Program), p. 157-166.

Dean, W.E., Arthur, M.A., and Claypool, G.E., 1986. Depletion of ^{13}C in Cretaceous marine organic matter: Source, diagenetic, or environment signal ? Mar. Geol., Vol. 70, p. 119-157.

Dean, W.E., Gardner, J.V., and Cepek, P., 1981. Tertiary carbonate-dissolution cycles on the Sierra Leone Rise, Eastern Equatorial Atlantic Ocean. Mar. Geol., Vol. 39, p. 81-101.

Degens, E.T., 1989. Perspectives on Biogeochemistry, Springer Verlag Berlin, 423 pp.

Degens, E.T. and Mopper, K., 1976. Factors controlling the distribution and early diagenesis of organic carbon in marine sediments. In: Riley, I.P. and Chester, R. (Eds.), Chem. Oceanogr., Vol. 6, London (Academic), p. 59-113.

Degens, E.T. and Ross, D.A., 1974. The Black Sea - Geology, Chemistry, and Biology. AAPG Memoir, Vol. 20, 633 pp.

De Graciansky, P.C., Deroo, G., Herbin, J.P., Montadert, L., Müller, C., Schaaf, A., and Sigal, J., 1984. Ocean-wide stagnation episode in the Late Cretaceous. Nature, Vol. 308, p. 346-349.

Demaison, G.J. and Moore, G.T., 1980. Anoxic environments and oil source bed genesis. Org. Geochem., Vol. 2, p. 9-31.

DeMenocal, P., Bristow. J., Stein, R., and Heusser, L., in prep. Paleoclimatic applications of logging data: Examples from Site 798. In: Ingle, J.C., Suyehiro, K., et al., Proc. ODP, Sci. Results, 128, College Station, Tx (Ocean Drilling Program).

Denton, G.H. and Amstrong, R.L., 1969 Miocene-Pliocene glaciations in southern Alaska. Amer. Journ. Sci., Vol. 267, p. 1121-1142.

Dersch, M., 1990. Zur Entwicklung von Paläoklima und paläoozeanischen Verhältnissen im SW-Pazifik: Ergebnisse von sedimentologischen Untersuchungen an Kernmaterial von DSDP-Site 594 (Chatham Rücken, östlich von Neuseeland). Univeröff. Diplomarbeit, Giessen University.

Dersch, M. and Stein, R., 1991. Paläoklima und paläoozeanische Verhältnisse im SW-Pazifik während der letzten 6 Mill. Jahre (DSDP-Site 594, Chatham Rücken, östl. Neuseeland). Geol. Rdschau, in press.

Dersch, M., and Stein, R., in prep. Late Cenozoic changes in clay mineral composition at ODP-Site 798 and their paleoenvironmental significance. In: Ingle, J.C., Suyehiro, K., et al., Proc. ODP, Sci. Results, 128, College Station, Tx (Ocean Drilling Program).

Deuser, W.G., 1987. Variability and hydrography and particle flux: transient and long-term relationships. In: Degens, E.T., et al. (Eds.), Particle Flux in the Ocean, Mitt. Geol. Pal. Inst., Univ. Hamburg, Vol. 62, p. 179-193.

De Vernal, A. and Mudie, P.J., 1989a. Late Pliocene to Holocene palynostratigraphy at ODP-Site 645, Baffin Bay. In: Srivastava, S.P., Arthur, M.A., et al., Proc. ODP, Sci. Results, 105, College Station, Tx (Ocean Drilling Program), p. 387-400.

De Vernal, A. and Mudie, P.J., 1989b. Pliocene and Pleistocene palynostratigraphy at ODP-Sites 646 and 647, Eastern and Southern Labrador Sea. In: Srivastava, S.P., Arthur, M.A., et al., Proc. ODP, Sci. Results, 105, College Station, Tx (Ocean Drilling Program), p. 401-422.

Diester-Haass, L., 1983. Differentiation of high oceanic fertility in marine sediments caused by coastal upwelling and/or river discharge off Northwest Africa during the Late Quaternary. In: Thiede, J. and Suess, E. (Eds.), Coastal Upwelling, Its Sediment Record, Part B, Plenum Press, New York, p. 399-419.

Dunbar, R., in prep. Pleistocene oxygen isotope stratigraphy of benthic and planktic foraminifers from the Japan Sea. In: Ingle, J.C., Suyehiro, K., et al., Proc. ODP, Sci. Results, 128, College Station, Tx (Ocean Drilling Program).

Dupont, L.M., Beug, H.J., Stalling, H., and Tiedemann, R., 1989. First palynological results from Site 658 at 21°N off Northwest Africa: Pollen as climate indicators. In: Ruddiman, W.F., Sarnthein, M., et al., Proc. ODP, Sci. Results, 108, College Station, Tx (Ocean Drilling Program), p. 93-112.

Durond, B., 1980. Kerogen. Editions Technip, Paris, 519 pp.

Ehrmann, W.U. and Thiede, J., 1985. History of Mesozoic and Cenozoic sediment fluxes to the North Atlantic Ocean. Contrib. Sedimentol., Vol. 15, 109 pp.

Ekdale, 1989. Trace fossils and original oxygen conditions in marine sediments. 28th Intern. Geol. Congr., Washington, Abstracts, Vol. 1, p. 438-439.

Eldholm, O., Thiede, J., et al., 1989. Proc. ODP Sci. Results, 104, College Station, Tx (Ocean Drilling Program), 1141 pp.

Emerson, S., 1985. Organic carbon preservation in marine sediments. In: Sundquist, E.T. and Broecker, W.S. (Eds.), The Carbon Cycle and Atmospheric CO_2: Natural Variations Archean to Present, Geophys. Monog., Vol. 32, p. 78-89.

Emerson, S., Fischer, K., Reimers, C., and Heggie, D., 1985. Organic carbon dynamics and preservation in deep-sea sediments. Deep-Sea Res., Vol. 32, p. 1-22.

Emerson, S., Stump, C., Grootes, P.M., Stuiver, M., Farwell, G.W., and Schmidt, F.G., 1987. Estimates of degradable organic carbon in deep-sea surface sediments from ^{14}C concentrations. Nature, Vol. 329, p. 51-53.

Emery, K.O. and Uchupy, E., 1984. The Geology of the Atlantic Ocean. Springer Verlag, New York, 1050 pp.

Eppley, R.W., 1989. New Production: History, Methods, Problems. In: Berger, W.H., et al. (Eds.), Productivity of the Ocean: Past and Present, Life Sci. Res. Rep., Vol. 44, Wiley & Sons, p. 85-98.

Eppley, R.W. and Peterson, B.J., 1979. Particulate organic matter flux and planktonic new production in the deep ocean. Nature, Vol. 282, p. 677-680.

Espitalié, J., Laporte, J.L., Madec, M., Marquis, F., Leplat, P., Paulet, J., and Boutefeu, A., 1977. Méthode rapide de characterisation des roches-mere, de leur potential petrolier et de leur degre d'évolution. Rev. Inst. Franc. Petrol., Vol. 32, p. 23-42.

Espitalié, J., Nakadi, K.S., and Trichet, J., 1984. Role of the mineral matrix during kerogen pyrolysis. Org. Geochem., Vol. 6, p. 365-382.

Faugères, J.C., Legigan, P., Maillet, N., and Latouche, C., 1989b. Pelagic, turbiditic, and contouritic sequential deposits on the Cape Verde Plateau (Leg 108, Site 659, Northwest Africa): Sediment record during neogene time. In: Ruddiman, W.F., Sarnthein, M., et al., Proc. ODP, Sci. Results, 108, College Station, Tx (Ocean Drilling Program), p. 311-328.

Faugères, J.C., Legigan, P., Maillet, N., Sarnthein, M., and Stein, R., 1989a. Characteristics and distribution of Neogene turbidites at Site 657 (Leg 108, Cap Blanc Continental Rise, Northwest Africa): Variations in turbidite source and continental climate. In: Ruddiman, W.F., Sarnthein, M., et al., Proc. ODP, Sci. Results, 108, College Station, Tx (Ocean Drilling Program), p. 329-349.

Fischer, G., Fütterer, D., Gersonde, R., Honjo, S., Ostermann, D., and Wefer, G., 1988. Seasonal variability of particle flux in the Weddell Sea and its relation to ice cover. Nature, Vol. 335, p. 426-428.

Föllmi, K.B., Alexandrovich, J., Brunner, C.A., Burckle, L.H., Charvet, J., Cramp, A., Demenocal, P., Dunbar, R.B., Grimm, K.A., Holler, P., Ingle, J.C., Isaacs, C.M., Kheradyar, T., T., Koizumi, I., Matsumoto, R., Nobes, D., Pisciotto, K., Rahnan, A., Stein, R., Tada, R., von Breymann, M., and Scientific Party of Legs 127 and 128, 1990. Paleoceanographic Implications from High-Frequency Dark/Light Rhythms in the Sea of Japan (ODP Legs 127 and 128). AAPG Circum Pacific Conference, 1990, Abstract.

Fontugne, M.R. and Duplessy, J.C., 1981. Organic carbon isotopic fractionation by marine plankton in the temperature range -1 to 31°C. Oceanologica Acta, Vol. 4, p. 85-89.

Frakes, L.A., 1979. Climates throughout geologic time. Elsevier Sci. Publ. Comp., Amsterdam, 310 pp.

Froelich, P.N., Klinkhammer, G.P., Bender, M.L., Luedtke, N.A., Heath, G.R., Collen, D., Dauphin, P., Hammond, D., and Hartman, B., 1979. Early oxidation of organic matter in pelagic sediments of the eastern equatorial Atlantic: suboxic diagenesis. Geochim. Cosmochim. Acta, Vol. 43, p. 1075-1090.

Gagosian, R.B. and Peltzer, E.T., 1987. The importance of atmospheric input of terrestrial organic material to deep sea sediments. Org. Geochem., Vol. 10, p. 661-669.

Ganssen, G. and Sarnthein, M., 1983. Stable-isotope composition of foraminifers: The surface and bottom water record of coastal upwelling. In: Suess, E. and Thiede, J. (Eds.), Coastal Upwelling: Its Sediment Record, Pt. A, Plenum Press, p. 99-121.

Genthon, C., Barnola, J.M., Raynaud, D., Lorius, C., Jouzel, J., Barkov, N.I., Korotkevich, Y.S., and Kotlyakov, V.M., 1987. Vostok ice core: climatic response to CO_2 and orbital forcing changes over the last climatic cycle. Nature, Vol. 329, p. 414-418.

Goldhaber, M.B. and Kaplan, I.R., 1974. The sulfur cycle. In: Goldberg, E.D. et al. (Eds.), The Sea Vol. 5, J. Wiley & Sons, New York, p. 569-655.

Günther, R. 1988. Rutschmassensedimente und ihr möglicher Ursprung am Kontinentalhang vor NW-Afrika (20°N) (ODP-Sites 657, 658). Unveröff. Diplomarbeit, Kiel University, 59 pp.

Hall, F.R., Bloemendal, J., King, J.W., Arthur, M.A., and Aksu, A.E., 1989. Middle to late Quaternary sediment fluxes in the Labrador Sea, ODP Leg 105, Site 646: A synthesis of rock-magnetic, oxygen-isotopic, carbonate, and planktonic foraminiferal data. In: Srivastava, S.P., Arthur, M.A., et al., Proc. ODP, Sci. Results, 105, College Station, Tx (Ocean Drilling Program), p. 653-688.

Hallam, A., 1984. Continental humid and arid zones during Jurassic and Cretaceous. Palaeogeogr., Palaeoclim., Palaeoecol., Vol. 47, p. 195-223.

Hallam, A., 1986. A review of Mesozoic climates. In: Summerhayes, C.P. and Shackleton, N.J. (Eds.), Proceedings of the "North Atlantic Paleoceanography" Conference, Geol. Soc. London, Spec. Publ., Vol. 21, p. 277-282.

Head, M.J., Norris, G., and Mudie, P.J., 1989a. Palynology and dinocyst stratigraphy of the Miocene in ODP Leg 105, Hole 645E, Baffin Bay. In: Srivastava, S.P., Arthur, M.A., et al., Proc. ODP, Sci. Results, 105, College Station, Tx (Ocean Drilling Program), p. 467-514.

Head, M.J., Norris, G., and Mudie, P.J., 1989b. Palynology and dinocyst stratigraphy of the upper Miocene and lowermost Pliocene, ODP-Leg 105, Site 646, Labrador Sea. In: Srivastava, S.P., Arthur, M.A., et al., Proc. ODP, Sci. Results, 105, College Station, Tx (Ocean Drilling Program), p. 423-452.

Heath, G.R., Moore, T.C., and Dauphin, J.P., 1977. Organic carbon in deep-sea sediments. In: Anderson, N.R. and Malahoff, A. (Eds.), The Fate of Fossil Fuel CO_2 in the Oceans. Plenum Press, New York, p. 605-625.

Hedges, J.I. and Mann, D.C., 1979. The lignin geochemistry of marine sediments from the southern Washington coast. Geochim. Cosmochim. Acta, Vol. 43, p. 1809-1818.

Heggie, D., Maris, C., Hudson, A., Dymond, J., Beach, R., and Cullen, J., 1987. Organic carbon oxidation and preservation in NW Atlantic continental marigin sediments. In: Geology and Geochemistry of Abyssal Plains, Geol. Soc. London, Spec. Publ., Vol. 31, p. 215-236.

Henrichs, S.M. and Reeburgh, W.S., 1987. Anaerobic mineralization of marine sediment organic matter: rates on the role of anaerobic processes in the oceanic carbon economy. Geomicrobiol. Journ., Vol. 5, p. 191-237.

Herbin, J.P., Montadert, L., Müller, C., Gomez, R., Thurow, J., and Wiedmann, J., 1986. Organic-rich sedimentation at the Cenomanian/Turonian boundary in oceanic and coastal basins in the North Atlantic and Tethys. In: Summerhayes, C.P. and Shackleton, N.J. (Eds.), Geol. Soc. London, Spec. Publ., Vol. 21, p. 389-422.

Hillaire-Marcel, C., de Vernal, A., Aksu, A.E., and Macko, S., 1989. High-resolution isotopic micropaleontological studies of upper Pleistocene sediments at ODP-Site 645, Baffin Bay. In: Srivastava, S.P., Arthur, M.A., et al., Proc. ODP, Sci. Results, 105, College Station, Tx (Ocean Drilling Program), 599-616.

Hiscott, R.N., Cremer, M., and Aksu, A.E., 1989. Evidence from sedimentary structures for processes of sediment transport and deposition during post-Miocene time at Sites 645, 646, and 647, Baffin Bay and Labrador Sea. In: Srivastava, S.P., Arthur, M.A., et al., Proc. ODP, Sci. Results, 105, College Station, Tx (Ocean Drilling Program), p. 53-64.

Hobart, M.A., Bunce, E.T., and Sclater, J.C., 1975. Bottom-water flow through the Kane Gap, Sierra Leone Rise, Atlantic Ocean. Journ. Geophys. Res., Vol. 80, p. 5083-5088.

Honjo, S., 1982. Seasonality and interaction of biogenic and lithogenic particle flux at the Panama Basin. Science, Vol. 218, p. 883-884.

Hughes, P. and Barton, E.D., 1974. Stratification and water mass structure in the upwelling area off Northwest Africa in April/May 1969. Deep-Sea Res., Vol. 21, p. 611-628.

Hutton, A.C., Kantsler, A.J., Cook, A.C., and McKirdy, D.M., 1980. Organic matter in oil shales. Austr. Petr. Expl. Ass., Vol. 20, p. 44-67.

Ibach, L.E., 1982. Relationships between sedimentation rate and total organic carbon content in ancient marine sediments. AAPG Bull., Vol. 66, p. 170-188.

Iijima, A., Tada, R., and Watanabe, W., 1988. Developments of Neogene sedimentary basins in the Northeastern Honshu Arc with emphasis on Miocene siliceous deposits. Journ. Facul. Sci., Tokyo, Vol. 21/5, p. 417-446.

Imbrie, J. and Kipp, N.G., 1971. A new micropaleontological method for quantitative paleoclimatology: Application to a late Pleistocene Caribbean core. In: Turekian, K.K. (Ed.), The Late Cenozoic glacial ages, New Haven, Yale Univ. Press, p. 71-181.

Ingle, J.C., 1975. Pleistocene and Pliocene foraminifera from the Japan Sea Leg 31, Deep Sea Drilling Project. In: Karig, D.E., Ingle, J.C., et al., Init. Reps. DSDP, Vol. 31, p. 693-701.

Ingle, J.C., 1981. Origin of Neogene diatomites around the north Pacific rim. In: Garrison, R.E., et al. (Eds.), The Monterey Formation and Related Siliceous Rocks of California, Pacific Sect. SEPM Spec. Publ., p. 159-179.

Ingle, J.C., Suyehiro, K., et al., 1990. Proc. ODP, Init. Repts., 128, College Station, Tx (Ocean Drilling Program), in press.

Isaacs, C.M., Pisciotto, K.A., and Garrison, R.E., 1983. Facies and diagenesis of the Monterey Formation, California: A summary. In: Iljima, A., et al. (Eds.), Developments in Sedimentology, Vol. 36, p. 247-281.

Ittekkot, V., 1988. Global trends in the nature of organic matter in river suspensions. Nature, Vol. 332, p. 436-438.

Izdar, E., Konuk, T., Ittekkot, V., Kempe, S., and Degens, E.T., 1983. Particle Flux in the Black Sea: Nature of the Organic Matter. In: Degens, E.T., et al. (Eds.), Particle Flux in the Ocean, Mitt. Geol. Pal. Inst., Hamburg University, Vol. 62, p. 1-18.

Jacobi, R. and Hayes, D., 1982. Bathymetry, microphysiography, and reflectivity characteristics of the West African margin between Sierra Leone and Mauritania. In: von Rad, U., et al. (Eds.), Geology of the Northwest African Continental Margin, Springer Verlag Berlin, p. 182-212.

Jansen, E., Sjoholm, J., Bleil, U., and Erichsen, J.A., 1990. Neogene and Pleistocene glaciations in the northern hemisphere and Miocene-Pliocene global ice volume fluctuations: Evidence from the Norwegian Sea. In: Bleil, U. and Thiede, J. (Eds.), Geological history of the Polar Oceans: Arctic versus Antarctic, Nato ASI Series C, Vol. 308, Kluwer Acad. Publ., p. 677-705.

Jasper, J.P. and Gagosian, R.B., 1989. Glacial-interglacial climatically forced $\delta^{13}C$ variations in sedimentary organic matter. Nature, Vol. 342, p. 60-62.

Jenkyns, H.C., 1985. The early Toarcian and Cenomanian-Turonian anoxic events in Europe: comparisons and contrasts. Geol. Rdsch., Vol. 74, p. 505-518.

Jones, R.W., 1983. Organic matter characteristics near the shelf-slope boundary. SEPM Spec. Publ., Vol. 33, p. 391-405.

Jumars, P.A., Altenbach, A.V., de Lange, G.J., Emerson, S.R., Hargrave, B.T., Müller, P.J., Prahl, F.G., Reimers, C.E., Steiger, T., and Suess, E., 1989. Transformation of Seafloor-arriving Fluxes into the Sedimentary Record. In: Berger, W.H., et al. (Eds.), Productivity of the Ocean: Past and Present, Life Sci. Res. Rep., Vol. 44, Wiley & Sons, p. 291-312.

Kastens, K.A., Mascle, J., et al., 1988. Proceedings Initial Reports A, ODP 107.

Katz, B.J., 1983. Limitations of 'Rock-Eval' pyrolysis for typing organic matter. Org. Geochim., Vol. 4, p. 195-199.

Keigwin, L.D., 1979. Late Cenozoic stable isotope stratigraphy and paleoceanography of DSDP sites from the East Equatorial and North Central Pacific Ocean. Earth Plan. Sci. Lett., Vol. 45, p. 361-382.

Kennett, J.P., Houtz, R.E., Andrews, P.B., Edwards, A.R., Gostin, V.A., Hajos, M., Hampton, M., Jenkins, D.G., Margolis, S.V., Ovenshine, A.T., and Perch-Nielsen, K., 1975. Cenozoic paleoceanography in the southwest Pacific Ocean, Antarctic glaciation, and the development of the Circum-Antarctic Current. In: Kennett, J.P., Houtz, R.E., et al., Init. Reps. DSDP, Vol. 29, p. 1155-1169.

Kettler, R., Rullkötter, J., Stein, R., and Kheradyar, T., in prep. Lipid geochemistry of diatomaceous sediments, ODP-Leg 128. In: Ingle, J.C., Suyehiro, K., et al., Proc. ODP, Sci. Results, 128, College Station, Tx (Ocean Drilling Program).

Kidd, R.B., Cita, M.B., and Ryan, W.B.F., 1978. Stratigraphy of eastern Mediterranean sapropel sequences recovered during DSDP Leg 42A and their paleoenvironmental significance. In: Hsü, K.J., Montadert, L., et al., Init. Reps. DSDP, Vol. 42, p. 421-443.

Knüttel, S., Russell, M.D., and Firth, J.V., 1989. Neogene calcareous nannofossils from ODP-Leg 105: Implications for Pleistocene paleoceanographic trends. In: Srivastava, S.P., Arthur, M.A., et al., Proc. ODP, Sci. Results, 105, College Station, Tx (Ocean Drilling Program), p. 245-262.

Kobayashi, K. and Nomura, M., 1972. Iron sulfides in the core from Sea of Japan and their geophysical implications. Earth Plan. Sci. Lett., Vol. 16, p. 200-208.

Koblents-Mishke, O.I., Volkovinsky, V.V., and Kabanova, Y.G., 1970. Plankton primary production of the World Ocean. In: Wooster, W. (Ed.), Scientific Exploration of the South Pacific, Nat. Acad. Sci., Washington, p. 183-193.

Kögler, F.C. and Larson, B., 1979. The West Bornholm Basin in the Baltic Sea: Geological structure and Quaternary sediments. Boreas, Vol. 8, p. 1-22.

Koopmann, B., 1981. Sedimentation von Saharastaub im subtropischen Nordatlantik während der letzten 25,000 Jahre. 'Meteor'-Forsch. Ergebn., C, Vol. 35, p. 23-59.

Korstgärd, J.A. and Nielsen, O.B., 1989. Provenance of dropstones in Baffin Bay and Labrador Sea, Leg 105. In: Srivastava, S.P., Arthur, M.A., et al., Proc. ODP, Sci. Results, 105, College Station, Tx (Ocean Drilling Program), p. 65-70.

Kulick, J., Leifeld, D.,Meisl, S., Pöschl, W., Stellmacher, R., Strecker, G., Theuerjahr, A.K., and Wolf, M., 1984. Petrofazielle und chemische Erkundung des Kupferschiefers der Hessischen Senke und des Harz-Westrandes. Geol. Jb, Reihe D, Vol. 68, 226 pp.

Lazarus, D. and Pallant, A., 1989. Oligocene and Neogene radiolarians from the Labrador Sea, ODP-Leg 105. In: Srivastava, S.P., Arthur, M.A., et al., Proc. ODP, Sci. Results, 105, College Station, Tx (Ocean Drilling Program), p. 349-380.

Lea, D.W., and Boyle, E.A., 1990. A 210,000-year record of barium variability in the deep northwest Atlantic Ocean. Nature, Vol. 347, p. 269-272.

Leinen, M. and Heath, G.R., 1982. Sedimentary indicators of atmospheric activity in the Northern Hemisphere during the Cenozoic. Palaeogeogr., Palaeoclim., Palaeoecol., Vol. 36, p. 1-36.

Leventhal, J.S., 1983. An interpretation of carbon and sulfur relationships in Black Sea sediments as indicators of environments of deposition. Geochim. Cosmochim. Acta, Vol. 47, p. 133-137.

Littke, R., 1987. Petrology and genesis of Upper Carboniferous seams from the Ruhr region, West Germany. Int. Journ. Coal Geol., Vol. 7, p. 147-184.

Lonsdale, P., 1978. Bedforms and the benthic boundary layer in the North Atlantic: a cruise report of *Indomed* Leg 11. SIO Reference, 78-30.

Lutze, G.F. and Coulbourn, W.T., 1984. Recent benthic foraminifera from the continental margin of Northwest Africa: community structure and distribution. Mar. Micropal., Vol. 8, p. 361-401.

Macko, S.A., 1989. Stable isotope organic geochemistry of sediments from the Labrador Sea (Sites 646 and 647) and Baffin Bay (Site 645), ODP Leg 105. In: Srivastava, S.P., Arthur, M.A., et al., Proc. ODP, Sci. Results, 105, College Station, Tx (Ocean Drilling Program), p. 209-232.

Marko, J.R., Birch, J.R., and Wilson, M.A., 1982. A study of long-term satellite-tracked iceberg drifts in Baffin Bay and Davis Strait. Arctic, Vol. 35, p. 234-240.

Marlowe, I.T., Green, J.C., Neal, A.C., Brassell, S.C., Eglinton, G., and Course, P.A., 1984. Long chain alkenones in the Prymnesiophyceae. Distribution of alkenones and other lipids and their taxonomic significance. Br. J. Phycol., Vol. 19, p. 203-216.

Matoba, Y., 1984. Paleoenvironment of the Sea of Japan. In: Benthos '83, Pau, Elf Aquitaine, p. 409-414.

McCave, I.N. and Tucholke, B.E., 1986. Deep Current-controlled Sedimentation in the Western North Atlantic. Geol. North Amer., Vol. M, The Western North Atlantic Region, Geol. Soc. Amer., p. 451-468.

Mienert, J., 1986. Akustostratigraphie im äquatorialen Ostatlantik: Zur Entwicklung der Tiefenwasserzirkulation der letzten 3.5 Mill. Jahre. 'Meteor'-Forsch. Ergebn., C, Vol. 40, p. 19-86.

Mienert, J., Stein, R., Schultheiss, P., and Shipboard Scientific Party, 1988. Relationship between grain density and biogenic opal in sediments from Sites 658 and 660. In: Ruddiman, W., Sarnthein, M., et al. (Eds.) Proceedings of the Ocean Drilling Program, Vol. 108, Part A, p. 1047-1053.

Miller, K.G. and Tucholke, B.E., 1983. Development of Cenozoic Abyssal Circulation South of the Greenland-Scotland Ridge. In: Bott, M.H.P., et al. (Eds.), Structure and development of the Greenland-Scotland Ridge, Plenum Press, New York, p. 549-590.

Mix, A.C., 1989. Pleistocene Paleoproductivity: Evidence from Organic Carbon and Foraminiferal Species. In: Berger, W.H., et al. (Eds.), Productivity of the Ocean: Past and Present, Life Sci. Res. Rep., Vol. 44, Wiley & Sons, p. 313-340.

Monjanel, A.L. and Baldauf, J., 1989. Miocene to Holocene diatom biostratigraphy from Baffin Bay and Labrador Sea, Ocean Drilling Program Sites 645 and 646. In: Srivastava, S.P., Arthur, M.A., et al., Proc. ODP, Sci. Results, 105, College Station, Tx (Ocean Drilling Program), p. 305-322.

Morris, R.J., McCartney, M.J., and Weaver, P.P.E., 1984. Sapropelic deposits in a sediment from the Guinea Basin, South Atlantic. Nature, Vol. 309, p. 611-614.

Mudie, P.J. and Helgason, J., 1983. Palynological evidence for Miocene climatic cooling in eastern Iceland about 9.8 Myr ago. Nature, Vol. 303, p. 689-692.

Mudie, P.J. and Short, S.K., 1985. Marine palynology of Baffin Bay. In: Andrews, J.T. (Ed.), Quaternary Environments, eastern Canadian Arctic, Baffin Bay, and western Greenland. Allen and Unwin, Boston, p. 263-308.

Müller, P.J., 1977. C/N ratios in Pacific deep-sea sediments: Effect of inorganic ammonium and organic nitrogen compounds sorbed by clays. Geochim. Cosmochim. Acta, Vol. 41, p. 765-776.

Müller, P., Erlenkeuser, H., and von Grafenstein, R., 1983. Glacial-interglacial cycles in oceanic productivity inferred from organic-carbon contents in eastern North Atlantic sediment cores. In: Thiede, J. and Suess, E. (Eds.), Coastal Upwelling: Its Sediment Record, Part B, Plenum Press, p. 365-398.

Müller, P.J. and Suess, E., 1979. Productivity, sedimentation rate, and sedimentary organic matter in the oceans. I.- Organic matter preservation. Deep-Sea Res., Vol. 26, p. 1347-1362.

Nelson, D.M., Smith, W.O., Muench, R.D., Gordon, L.I., Sullivan, C.W., and Husby, D.M., 1989. Particulate matter and nutrient distribution in the ice-edge zone of the Weddell Sea: relationship to hydrography during late summer. Deep-Sea Res., Vol. 36, p. 191-209.

Niedermeyer, R.O. and Lange, D., 1990. An Actualistic Model of Mud deposition and Diagenesis for the Western Baltic Sea. Limnologica, Vol. 20, p. 9-14.

Oba, T., 1983. Oxygen isotope analysis. In: Kobayashi, K. (Ed.), Preliminary Report of the Hakuho-maru Cruise KH 82-4, Tokyo, p. 140-141.

Oba, T., Kato, M., Kitazato, H., Koizumi, I., Omura, A., Sakai, T., and Takayama, T., 1990. Paleoenvironmental changes in the Japan Sea during the last 85,000 years. Paleoceanography, in press.

Olausson, E., 1961. Studies of deep-sea cores. Rep. Swedish Deep-Sea Exped., Vol. 8, p. 337-391.

Orr, W.L., 1983. Comments on Pyrolytic Hydrocarbon Yields in Source-Rock Evaluation. Advances in Organic Geochemistry 1981, Wiley & Sons, p. 775-787.

Osterman, L.E., 1982. Late Quaternary history of southern Baffin Island, Canada: a study of foraminifera and sediments from Frobisher Bay. PhD thesis, University of Colorado, Boulder, 380 pp.

Paul, J., 1982. Zur Rand- und Schwellenfazies des Kupferschiefers. Zt. dt. geol. Ges., Vol. 153, p. 571-605.

Pelet, R. and Debyser, Y., 1977. Organic geochemistry of Black Sea cores. Geochim. Cosmochim. Acta, Vol. 41, p. 155-1586.

Pelletier, B.R., Ross, D.I., Keen, C.E., and Keen, M.J., 1975. Geology and Geophysics of Baffin Bay. Geol. Surv. Canada, Paper 74-30, p. 247-258.

Pisciotto, K., Tamaki, K., et al., 1990. Proc. ODP, Init. Repts., 127: College Station, Tx (Ocean Drilling Program), in press.

Pocklington, R., 1976. Terrigenous Organic Matter in Surface Sediments from the Gulf of St. Lawrence. Journ. Fish. Res. Board Can., Vol. 33, p. 93-97.

Postma, H., 1982. Hydrography of the Wadden Sea: Movements and properties of water and particulate matter. Balkema, Rotterdam, 75 pp.

Poutanen, E.L. and Morris, R.J., 1985. Humic substances in an Arabian shelf sediment and the S1 sapropel from the eastern Mediterranean. Chem. Geol., Vol. 51, p. 135-145.

Prahl, F.G., de Lange, G.J., Lyle, M., and Sparrow, M.A., 1989. Post-depositional stability of long-chain alkenones under contrasting redox conditions. Nature, Vol. 341, p. 434-437.

Prahl, F.G. and Carpenter, R., 1984. Hydrocarbons in Washington coastal sediments. Est. Coast. Shelf Sci., Vol. 18, p. 703-720.

Prahl, F.G. and Muehlhausen, L.A., 1989. Lipid Biomarkers as Geochemical Tools for Paleoceanographic Study. In: Berger, W.H., et al. (Eds.), Productivity of the Ocean: Past and Present, Life Sci. Res. Rep., Vol. 44, Wiley & Sons, p. 271-290.

Pratt, L.M., 1984. Influence of paleoenvironmental factors on preservation of organic matter in Middle Cretaceous Greenhorn formation, Pueblo, Colorado. AAPG Bull., Vol. 68, p. 1146-1159.

Prell, W.L., Niitsumo, N. et al. 1988. Proc. ODP, Init. Repts., 117, College Station, Tx (Ocean Drilling Program), 1236 p.

Premuzic, E.T., Benkovitz, C.M., Gaffney, J.S., and Walsh, J.J., 1982. The nature and distribution of organic matter in the surface sediments of world oceans and seas. Org. Geochem., Vol. 4, p. 63-77.

Raiswell, R. and Berner, R.A., 1987. Organic carbon losses during burial and thermal maturation of normal marine shales. Geology, Vol. 15, p. 853-856.

Reimers, C.E., 1989. Control of Benthic Fluxes by Particulate Supply. In: Berger, W.H., et al. (Eds.), Productivity of the Ocean: Past and Present, Life Sci. Res. Rep., Vol. 44, Wiley & Sons, p. 217-234.

Reimers, C.E. and Suess, E., 1983. Late Quaternary fluctuations in the cycling of organic matter off central Peru: A protokerogen record. In: Suess, E. and Thiede, J. (Eds.), Coastal Upwelling, Its Sediment Record, Part A, Pelnum Press, New York, p. 497-526.

Ricken, W., 1989. Independent Measurement of Time and Sediment Flux in Bedding Cycles and Larger Sequences. 28th Int. Geol. Congr., Washington, Abstarcts, Vol. 2, p. 696-697.

Riegel, W., Loh, H., Maul, B., and Prauss, M., 1986. Effects and causes in a black shale event - the Toarcian Posidonia Shale of NW Germany. In: Walliser, O.H. (Ed.), Global Bio-Events, Lect. Not. Earth Sci., Vol. 8, p. 267-276.

Roberts, Schnitker, et al. 1984. Init. Reps. DSDP, Vol. 81, 923 pp.

Romankevich, E.A., 1984. Geochemistry of Organic Matter in the Ocean. Springer Verlag Berlin, 334 pp.

Ross, D.A., Degens, E.T., MacIlvaine, J., 1970. Black Sea: recent sedimentary history. Science, Vol. 170, p. 163-165.

Rossignol-Strick, M., Nesteroff, W., Olive, P., and Vergnaud-Grazzini, C., 1982. After the deluge: Mediterranean stagnation and sapropel formation. Nature, Vol. 295, p. 105-110.

Rother, K., 1989. Petro- und paläomagnetische Untersuchungen an jungquartären Sedimenten der Ostsee. Veröff. Zentr. Inst. Physik der Erde, Vol. 109, 229 pp.

Ruddiman, W., Sarnthein, M., et al., 1988. Proc. ODP, Init. Repts., 108, College Station, Tx (Ocean Drilling Program), 556 pp.

Ruddiman, W.F. and Janecek, T.R., 1989. Pliocene-Pleistocene biogenic and terrigenous fluxes at Equatorial Atlantic Sites 662, 663, and 664. In: Ruddiman, W.F., Sarnthein, M., et al., Proc. ODP, Sci. Results, 108, College Station, Tx (Ocean Drilling Program), p. 211-240.

Ruddiman, W.F., Sarnthein, M., Backman, J., Baldauf, J.G., Curry, W., Dupont, L.M., Janecek, T.R., Pokras, E.M., Raymo, M.E., Stabell, B., Stein, R., and Tiedemann, R., 1989. Late Miocene to Pleistocene evolution of climate in Africa and the low-latitude Atlantic: Overview of Leg 108 results. In: Ruddiman, W.F., Sarnthein, M., et al., Proc. ODP, Sci. Results, 108, College Station, Tx (Ocean Drilling Program), 463-486.

Rullkötter, J., Mukhopadhyay, P.K., Schaefer, R.G., and Welte, D.H., 1984. Geochemistry and petrography of organic matter in sediments from DSDP Sites 545 and 547, Mazagan Escarpment. In: Hinz, K., Winterer, E.L., et al., Init. Reps. DSDP, Vol 79, p. 775-806.

Rullkötter, J., Vuchev, V., Hinz, K., Winterer, E.L., Bradshaw, M.J., Channell, J.E.T., Jaffrezo, M., Jansa, L.F., Leckie, R.M., Moore, J.M., Schaftenaar, C., Steiger, T.H., and Wiegand, G.E., 1983. Potential deep-sea petroleum source beds related to coastal upwelling. In: Thiede, J. and Suess, E. (eds.), Coastal Upwelling, Its Sediment Record, Part B, Plenum Press, New York, p. 467-483.

Sackett, W.M. and Thompson, R.R., 1963. Bull. Amer. Assoc. Petrol. Geol., Vol.47, p. 525.

Sakshaug, E. and Holm-Hansen, O., 1984. Factors Governing Pelagic Production in Polar Oceans. In: Holm-Hansen, O., et al. (Eds.), Marine Phytoplankton and Productivity, Lect. Notes Coast. Est. Studies, Vol. 8, p. 1-17.

Sandberg, C.A., Ziegler, W., Dreesen, R., and Butler, J.L., 1988. Late Frasnian Mass Extinction: Conodont Event Stratigraphy, Global Changes, and possible Causes. Cour. Forsch.-Inst. Senckenberg, Vol. 102, p. 263-307.

Sarnthein, M., Tetzlaff, G., Koopmann, B., Wolter, K., and Pflaumann, U., 1981. Glacial and interglacial wind regimes over the eastern subtropical Atlantic and Northwest Africa. Nature, Vol. 293, p. 193-196.

Sarnthein, M., Thiede, J., Pflaumann, U., Erlenkeuser, H., Fütterer, D., Koopmann, B., Lange, H., and Seibold, E., 1982. Atmospheric and oceanic circulation patterns off

Northwest Africa during the past 25 million years. In: von Rad, U., et al. (Eds.), Geology of the Northwest African Continental Margin, Springer Verlag Berlin, p. 545-604.

Sarnthein, M. and Tiedemann, R., 1989. Toward a high-resolution stable isotope stratigraphy of the last 3.4 Million years: Sites 658 and 659 off Northwest Africa. In: Ruddiman, W.F., Sarnthein, M., et al., Proc. ODP, Sci. Results, 108, College Station, Tx (Ocean Drilling Program), p. 167-186.

Sarnthein, M., Winn, K., Duplessy, J.C., and Fontugne, M.R., 1988. Global variations of surface water productivity in low- and mid-latitudes: influence on CO_2 reservoirs of the deep ocean and atmosphere during the last 21,000 years. Paleoceanography, Vol. 3, p. 361-399.

Sarnthein, M., Winn, K., and Zahn, R., 1987. Paleoproductivity of oceanic upwelling and the effect on atmospheric CO_2 and climatic change during deglaciation times. In: Berger, W.H. and Labeyrie, L. (Eds.), Abrupt Climatic Change, Riedel Publ., Dordrecht, p. 311-337.

Schaeffer, R. and Spiegler, D., 1986. Neogene Kälteeinbrüche und Vereisungsphasen im Nordatlantik. Z. Deutsch. Geol. Ges., Vol. 137, p. 537-552.

Scheffer, F. and Schachtschabel, P., 1984. Lehrbuch der Bodenkunde. F. Enke, Stuttgart, 442 pp.

Scheidgger, K.F. and Krissek, L.A., 1983. Zooplankton and nekton: Natrural barriers to the seaward transport of suspended terrigenous particles off Peru. In: Suess, E. and Thiede, J. (Eds.), Coastal Upwelling, Its Sediment Record, Part A, Plenum Press, p. 303-336.

Schemainda, R., Nehring, D., and Schulz, S., 1975. Ozeanologische Untersuchungen zum Produktionspotential der nordwestafrikanischen Wasserauftriebsregion 1970-1973. Geodätische und Geophysikalische Veröff., Reihe IV, Vol. 16, 85 pp.

Schlanger, S.O. and Jenkyns, H.C., 1976. Cretaceous oceanic anoxic events - causes and consequences. Geologie en Mijnbouw, Vol. 55, p. 179-184.

Schmitz, B., 1987. Barium, equatorial high-productivity and the northward wandering of the Indian continent. Paleoceanography, Vol. 2, p. 63-77.

Schrader, H. and Matherne, A., 1981. Sapropel formation in the eastern Mediterranean: evidence from preserved opal assemblages. Micropal., Vol. 27, p. 191-203.

Shackleton, N.J. and Opdyke, N.D., 1977. Oxygen isotope and paleomagnetic evidence for early Northern Hemisphere Glaciation. Nature, Vol. 270, p. 216-219.

Shackleton, N.J. and Pisias, N.G., 1985. Atmospheric carbon dioxide, orbital forcing, and climate. In: Sundquist, E.T. and Broecker, W.S. (Eds.), The Carbon Cycle and Atmospheric CO_2: Natural Variations Archean to Present, Geophys. Monog., Vol. 32, p. 303-317.

Shackleton, N.J., Backman, J., Zimmerman, H., Kent, D.V., Hall, M.A., Roberts, D.G., Schnitker, D., Baldauf, J.G., Desprairies, A., Homrighausen, R., Huddlestun, P., Keene, J.B., Kaltenback, A.J., Krumsiek, K.A.O., Morton, A.C., Murray, J.W., and Westberg-Smith, J., 1984. Oxygen isotope calibration of the onset of ice-rafting and history of glaciation in the North Atlantic region. Nature, Vol. 37, p. 620-623.

Shaffer, G., 1976. A mesoscale study of coastal upwelling variability off NW-Africa. 'Meteor'-Forsch. Ergebn., A, Vol. 17, p. 33-70.

Shimkus, K.M. and Trimonis, E.S., 1974. Modern sedimentation in the Black Sea. In: Degens, E.T. and Ross, D.A. (Eds.), The Black Sea - Geology, Chemistry, and Biology, AAPG Mem., Vol. 20, p. 249-278.

Siegenthaler, U. and Wenk, T., 1984. Rapid atmospheric CO_2 variations and ocean circulation. Nature, Vol. 308, p. 624-626.

Sigl, W., Chamley, H., Fabricius, F., d'Argoud, G.G., and Müller, J., 1978. Sedimentology and environmental conditions of sapropels. In: Hsü, K.J., Montadert, L., et al., Init. Reps. DSDP, Vol. 42, p. 445-465.

Sliter, W.V., 1989. Aptian anoxia in the Pacific Basin. Geology, Vol. 17, p. 909-912.

Smith, B.N., 1976. Evolution of C4 photosynthesis in response to changes in carbon and oxygen concentrations in the atmosphere through time. BioSystems, Vol. 8, p. 24-32.

Smith, R.L., 1983. Circulation patterns in upwelling regimes. In: Suess, E., and Thiede, J. (Eds.), Coastal Upwelling, Its Sediment Record, Part A, Plenum Press, New York, p. 13-35.

Sorokin, Y.I., 1982. Black Sea. Nauka, Moscow, 216 pp.

Srivastava, S.P., Arthur, M.A., et al., 1987. Proc. ODP, Init. Repts., 105, College Station, Tx (Ocean Drilling Program), 917 p.

Srivastava, S., Falconer, R.K.H., and MacLean, B., 1981. Labrador Sea, Davis Strait, Baffin Bay: Geology and Geophysics - A Review. In: Kerr, J.W., et al. (Eds.), Geology of the North Atlantic Borderlands. Can. Soc. Petr. Geol. Memoir 8, p. 333-398.

Stabell, B., 1989. Initial diatom record of Sites 657 and 658: On the history of upwelling and continental aridity. In: Ruddiman, W.F., Sarnthein, M., et al., Proc. ODP, 108, College Station, Tx (Ocean Drilling Program), p. 149-156.

Stach, E., Mackowsky, M.T., Teichmüller, M., Taylor, G.H., Chandra, D., and Teichmüller, R., 1982. Stach's Textbook of Coal Petrology. Borntraeger Verlag, Berlin, 535 pp.

Stanley, D.J., 1978. Ionian Sea sapropel distribution and late Quaternary paleoceanography in the eastern Mediterranean. Nature, Vol. 274, p. 149-152.

Stax, R., 1989. Zur Entwicklung von Paläoklima und paläoozeanischen Verhältnissen in der Labrador See: Ergebnisse aus sedimentologischen Untersuchungen an känozoischen Sedimenten von ODP-Site 647. Unveröff. Diplomarbeit, Universität Giessen, 66 pp.

Stax, R. and Stein, R., 1989. Late Plio-Pleistocene changes in sedimentary facies at Labrador-Sea Site 647 and their paleoenvironmental significance, Terra, Vol. 1, p. 440.

Stein, R. 1984. Zur neogenen Klimaentwicklung in Nordwest- Afrika und Paläo-Ozeanographie im Nordost- Atlantik. Berichte- Reports, Geol. Pal. Inst. Univ. Kiel, Nr. 4, 210 S. (Dissertation).

Stein, R., 1985a. Late Neogene Changes of Paleoclimate and Paleoproductivity off Northwest Africa (DSDP- Site 397). Palaeogeogr., Palaeoclim., Palaeoecol., 49, p. 47-59.

Stein, R., 1985b. The Post- Eocene Sediment Record of DSDP- Site 316: Implications for African Climate and Plate Tectonic Drift. Geol. Soc. Amer., Memoir 163, p.305-315.

Stein, R., 1986a. Surface-water paleo-productivity as inferred from sediments deposited in oxic and anoxic deep-water environments of the Mesozoic Atlantic Ocean. In: Degens, E.T., et al. (Eds.), Biochemistry of Black Shales, Mitt. Geol. Paläont. Inst. Univ. Hamburg, Vol. 60, p. 55-70.

Stein, R., 1986b. Organic carbon and sedimentation rate - Further evidence for anoxic deep-water conditions in the Cenomanian/Turonian Atlantic Ocean. Mar. Geol., Vol. 72, p. 199-209.

Stein, R., 1986c. Late Neogene evolution of paleoclimate and paleoceanic circulation in the Northern and Southern Hemisphere - A comparison. Geol. Rdschau, Vol. 75, p. 125-138.

Stein, R., 1990. Organic carbon content/sedimentation rate relationship and its paleoenvironmental significance for marine sediments. Geo-Mar. Lett., Vol. 10, p. 37-44.

Stein, R., 1991. Organic carbon accumulation in Baffin Bay and paleoenvironment in High-Northern Latitudes during the past 20 m.y. Geology, in press.

Stein, R., and Dunbar, R., in prep. Pliocene-Pleistocene fluctuations in eolian sediment supply at Site 798 (Oki Ridge, Sea of Japan) and climatic change. In: Ingle, J.C., Suyehiro, K., et al., Proc. ODP, Sci. Results, 128, College Station, Tx (Ocean Drilling Program).

Stein, R. and Faugères, J.C., 1989. Sedimentological and geochemical characteristics of the upper Cretaceous and lower Tertiary sediments at Site 661 (Eastern Equatorial Atlantic) and their paleoenvironmental significance. In: Ruddiman, W.F., Sarnthein, M., et al., Proc. ODP, Sci. Results, 108, College Station, Tx (Ocean Drilling Program), p. 297-310.

Stein, R. and Littke, R., 1990. Organic-carbon-rich sediments and paleoenvironment: Results from Baffin Bay (ODP-Leg 105) and the upwelling area off Northwest Africa (ODP-Leg 108). In: Huc, A. (Ed.), Deposition of Organic Facies, AAPG Studies in Geology, Vol. 30, p. 41-56.

Stein, R., Littke, R., Stax, R., and Welte, D.H., 1989a. Quantity, provenance, and maturity of organic matter at ODP-Sites 645, 646, and 647: Implications for reconstruction of paleoenvironments in Baffin Bay and Labrador Sea during Tertiary and Quaternary time. In: Srivastava, S.P., Arthur, M.A., et al., Proc. ODP, Sci. Results, 105, College Station, Tx (Ocean Drilling Program), p. 185-208.

Stein, R., Rullkötter, J., and Welte, D.H., 1986. Accumulation of organic-carbon-rich sediments in the Late Jurassic and Cretaceous Atlantic Ocean - A synthesis. Chem. Geol., Vol. 56, p. 1-32.

Stein, R., Rullkötter, J., and Welte, D.H., 1989c. Changes in paleoenvironments in the Atlantic Ocean during Cretaceous times: results from black shales studies. Geol. Rdsch., Vol. 78, p. 883-901.

Stein, R., Rullkötter, J., Littke, R., Schaefer, R. G., and Welte, D. H., 1988. Organofacies reconstruction and lipid geochemistry of sediments from the Galicia

Margin, Northeast Atlantic (ODP-Leg 103). In: Boillot, Winterer, et al. (Eds.), Proc. ODP, Sci. Results, 103, College Station, Tx (Ocean Drilling Program), p. 567-585.

Stein, R. and Sarnthein, M., 1984. Late Neogene events of atmospheric and oceanic circulation offshore Northwest Africa: High- resolution record from deep-sea sediments. Paleoecology of Africa, Vol. 16, p. 9-36.

Stein, R. and Stax, R., 1991. Late Quaternary organic carbon cycles and paleoenvironment in the northern Labrador Sea (ODP-Site 646). Geo-Mar. Lett., in press.

Stein, R., and Stax, R., in prep. Short- and long-term variations in flux rates and composition of organic carbon at Sites 798 and 799 through late Cenozoic times: Implications for basin evolution, paleoclimate, and paleoceanography in the Sea of Japan. In: Ingle, J.C., Suyehiro, K., et al., Proc. ODP, Sci. Results, 128, College Station, Tx (Ocean Drilling Program).

Stein, R., ten Haven, H.L., Littke, R., Rullkötter, J., and Welte, D.H., 1989b. Accumulation of marine and terrigenous organic carbon at upwelling Site 658 and nonupwelling Sites 657 and 659: Implications for the reconstruction of paleoenvironments in the eastern subtropical Atlantic through late Cenozoic times. In: Ruddiman, W.F., Sarnthein, M., et al., Proc. ODP, Sci. Results, 108, College Station, Tx (Ocean Drilling Program), p. 361-386.

Stevenson, F.J. and Cheng, C.N., 1972. Organic geochemistry of the Argentine Basin sediments: Carbon-nitrogen relationships and Quaternary correlations. Geochim. Cosmochim. Acta, Vol. 36, p. 653-671.

Suess, E., 1980. Particulate organic carbon flux in the oceans - surface productivity and oxygen utilisation. Nature, Vol. 288, p. 260-263.

Suess, E., von Huene, R. et al., 1988. Proc. ODP, Init. Repts., 112: College Station, Tx (Ocean Drilling Program), 1015 p.

Sutherland, H.E., Calvert, S.E., and Morris, R.J., 1984. Geochemical studies of the Recent sapropel and associated sediment from the Hellenic outer ridge, eastern Mediterranean Sea. I. Mineralogy and chemical composition. Mar. Geol., Vol. 56, p. 79-92.

Tada, R., Watanabe, Y., and Iijima, A., 1986. Accumulation of laminated and bioturbated Neogene siliceous deposits in Ajigasawa and Goshogawara areas, Aomori Prefecture, Northeast Japan. Journ. Facul. Sci., Tokyo, Vol. 21/3, p. 139-167.

Tamaki, K., 1985. Two modes of back-arc spreading. Geology, Vol. 13, p. 475-478.

Tamaki, K., 1988. Geological structure of the Japan Sea and its tectonic implications. Bull. Geol. Surv. Japan, Vol. 39, p. 269-365.

Tan, F.C. and Strain, P.M., 1979. Organic carbon isotope ratios in recent sediments in the St. Lawrence Estuary and the Gulf of St. Lawrence. Est. Coast. Mar. Sci., Vol. 8,213-225.

Tchernia, P., 1982. Descriptive Regional Oceanography. Pergamon Marine Series, Vol. 3, Pergamon, Oxford, 253 p.

Ten Haven, H.L., Baas, M., de Leeuw, J.W., Schenck, P.A., and Brinkhuis, H., 1987. Late Quaternary Mediterranean sapropels. II. Organic geochemistry and palynology of S1 sapropels and associated sediments. Chem. Geol., Vol. 64, p. 149-167.

Ten Haven, H.L. and Rullkötter, J., 1989. Oleanene, ursene, and other terrigenous triterpenoid biological-marker hydrocarbons in Baffin Bay sediments. In: Srivastava, S.P., Arthur, M.A., et al., Proc. ODP, Sci. Results, 105, College Station, Tx (Ocean Drilling Program), p. 233-243.

Ten Haven, H.L., Littke, R., Rullkötter, J., Stein, R., and Welte, D.H., 1990. Accumulation rates and composition of organic matter in late Cenozoic sediments underlying the active upwelling area off Peru. In: Suess, E., von Huene, R., et al., Proc. ODP, Sci. Results 112, College Station, Tx (Ocean Drilling Program), p. 591-606.

Ten Haven, H.L., Rullkötter, J., and Stein, R., 1989. Preliminary analysis of extractable lipids in sediments from the eastern North Atlantic (Leg 108): Comparison of a coastal upwelling area (Site 658) with a nonupwelling area (Site 659). In: Ruddiman, W.F., Sarnthein, M., et al., Proc. ODP, Sci. Results, 108, College Station, Tx (Ocean Drilling Program), p. 351-360.

Tetzlaff, G. and Wolter, W., 1980. Meteorological patterns and the transport of mineral dust from the North African continent. Palaeoecology of Africa, Vol. 12, p. 31-42.

Thein, J. and von Rad, U., 1987. Silica diagenesis in continental rise and slope sediments off eastern North America (Sites 603 and 605, Leg 93; Sites 612 and 614, Leg 95). In: Poag, C.W., Watts, A.B., et al., Init. Reps. DSDP, Vol. 95, p. 501-525.

Thiebault, F., Cremer, M., Debrabant, P., Foulon, J., Nielsen, O.B., and Zimmerman, H., 1989. Analysis of sedimentary facies, clay mineralogy, and geochemistry of the Neogene-Quaternary sediments in ODP-Site 645, Baffin Bay. In: Srivastava, S.P., Arthur, M.A., et al., Proc. ODP, Sci. Results, 105, College Station, Tx (Ocean Drilling Program), p. 83-100.

Thiede, J., 1979. Wind regimes over the late Quaternary southwest Pacific Ocean. Geology, Vol. 7, p. 259-262.

Thiede, J., 1983. Skeletal plankton and nekton in upwelling water masses off northwestern South America and Northwest Africa. In: Suess, E. and Thiede, J. (Eds.) Coastal Upwelling, Its Sediment Record, Part A, Plenum Press, New York, p. 183-207.

Thiede, J. and Eldholm, O., 1983. Speculations about the Paleodepth of the Greenland-Scotland Ridge during Late Mesozoic and Cenozoic Times. In: Bott, M.H.P., et al. (Eds.), Structure and development of the Greenland-Scotland Ridge, Plenum Press, New York, p 445-456.

Thiede, J. and van Andel, T.H., 1977. The paleoenvironment of anaerobic sediments in the Late Mesozoic South Atlantic Ocean. Earth Plan. Sci. Lett., Vol. 33, p. 301-309.

Thunell, R.C., 1986. Pliocene-Pleistocene climatic changes: evidence from land-based and deep-sea marine records. Mem. Soc. Geol. Ital., Vol. 31, p. 125-143.

Thunell, R.C. and Williams, D.F., 1989. Glacial-Holocene salinity changes in the Mediterranean Sea: hydrographic and depositional effects. Nature, Vol. 338, p. 493-496.

Thunell, R.C., Williams, D.F., and Belyea, P.R., 1984. Anoxic events in the Mediterranean Sea in relation to the evolution of the Late Neogene climates. Mar. Geol., Vol. 59, p. 105-134.

Thunell, R.C., Williams, D.F., and Kennett, J.P., 1977. Late Quaternary paleoclimatology, stratigraphy and sapropel history in eastern Mediterranean deep-sea sediments. Mar. Micropal., Vol. 2, p. 371-388.

Tiedemann, R., 1985. Verteilung von organischem Kohlenstoff in Oberflächensedimenten und die örtliche Primärproduktion im äquatorialen Ostatlantik, 0-20°N, 15-25°W. Unveröff. Diplomarbeit, Kiel University.

Tiedemann, R., Sarnthein, M., and Stein, R., 1989. Climatic changes in the western Sahara: Aeolo-marine sediment record of the last 8 million years (Sites 657-661). In: Ruddiman, W.F., Sarnthein, M., et al., Proc. ODP, Sci. Results, 108, College Station, Tx (Ocean Drilling Program), p. 241-278.

Tissot, B., Demaison, G., Masson, P., Delteil, J.R., and Combaz, A., 1980. Paleoenvironment and petroleum potential of mid-Cretaceous black shales in the Atlantic Basins. AAPG Bull., Vol. 64, p. 2051-2063.

Tissot, B., Deroo, G., and Herbin, J.P., 1979. Organic matter in Cretaceous sediments of the North Atlantic: Contribution to sedimentology and paleoceanography. In: Talwani, M., et al. (Eds.), Deep Drilling Results in the Atlantic Ocean: Continental Margin and Paleoenvironment, Amer. Geophys. Union, M. Ewing Ser., Vol. 3, p. 362-374.

Tissot, B.P. and Welte, D.H., 1984. Petroleum Formation and Occurrence. Springer Verlag Berlin, 699 pp.

Tomczak, M. and Hughes, P., 1980. Three dimensional variability of water masses and currents in the Canary Current upwelling region. 'Meteor'-Forsch. Ergebn., A, Voil. 21, p. 1-24.

Tucholke, B.E. and Mountain, G.S., 1986. Tertiary Paleoceanography of the Western North Atlantic Ocean. In: Vogt, P.R. and Tucholke, B.E. (Eds.), The Geology of North America, Vol. M, The Western Atlantic Region, Geol. Soc. Amer., p. 631-650.

Tungsheng, L., 1988. Loess in China. Springer Verlag, Berlin, 224 pp.

Ujiie, H. and Ichikura, M., 1973. Holocene to uppermost Pleistocene planktonic foraminifers in a piston core from off the San'n district, Sea of Japan. Trans. Palaeont. Soc. Japan, New Ser., Vol. 9, p. 137-150.

Valeton, I., 1983. Klimaperioden lateritischer Verwitterung und ihr Abbild in den synchronen Sedimentationsräumen. Z. dt. geol. Ges., Vol. 134, p. 413-452.

Van Andel, T.H., Heath, G.R., and Moore, T.C., 1975. Cenozoic History and Paleoceanography of the Central Equatorial Pacific Ocean. Geol. Soc. Amer., Memoir 143.

Vaynshteyn, M.B., Tokarev, V.G., Shakola, V.A., Lein, A.Y., and Ivanov, M.V., 1986. The geochemical activity of sulfate-reducing bacteria in sediments in the western part of the Black Sea. Geochem. Intern., Vol. 23, p. 110-122.

Vergnaud-Grazzini, C., Ryan, W.B.F., and Cita, M.B., 1977. Stable isotopic fractionation, climate change and episodic stagnation in the eastern Mediterranean during the late Quaternary. Mar. Micropal., Vol. 2, p. 353-370.

Vincent, E. and Berger, W.H., 1985. Carbon dioxide and polar cooling in the Miocene: The Montary hypothesis. In: Sundquist, E.T. and Broecker, W.S. (Eds.), The Carbon Cycle and Atmospheric CO_2: Natural Variations Archean to Present, AGU Geophys. Monogr., Vol. 32, p. 455-468.

Volkman, J.K., Eglinton, G., Corner, E.D.S., and O'Hara, S.C.M., 1980. Novel unsaturated straight chain C_{37}-C_{39} methyl and ethyl ketones in marine sediments and a coccolithophore Emiliania huxleyi. In: Douglas, A.G. and Maxwell, J.R. (Eds.), Advances in Organic Geochemistry 1979, Pergamon Press, New York, p. 219-227.

Von Breymann, M.T., Emeis, K.C., and Suess, E., 1991. Water depth and diagenetic constraints on the use of barium as a paleoproductivity indicator. In: Summerhayes, C.P., Prell, W.L., and Emeis, K.C. (Eds.), Proc. "Evolution of Upwelling Systems since the Early Miocene" Conference, London, September 1990, subm.

Walliser, O.H., 1986. Global Bio-Events. Lect. Not. Earth Sci., Vol. 8, 442 pp.

Walsh, J.J., 1989. How Much Shelf Production Reaches the Deep Sea ? In: Berger, W.H., et al. (Eds.), Productivity of the Ocean: Past and Present, Life Sci. Res. Rep., Vol. 44, Wiley & Sons, p. 175-192.

Waples, D., 1981. Organic Geochemistry for Exploration Geologists. Burgess Publ. Comp., Minnesota, 151 pp.

Waples, D.W., 1983. Reppraisal of anoxia and organic richness, with emphasis on Cretaceous of the North Atlantic. AAPG Bull., Vol. 61, p. 963-978.

Weaver, P.P.E. and Kuijpers, A., 1983. Climatic control of turbidites deposition on the Madeira Abyssal Plain. Nature, Vol. 306, p. 360-363.

Weaver, P.P.E. and Raymo, M.E., 1989. Late Miocene to Holocene planktonic foraminifers from the equatorial Atlantic, Leg 108. In: Ruddiman, W.F., Sarnthein, M., et al., Proc. ODP, Sci. Results, 108, College Station, Tx (Ocean Drilling Program), p. 71-92.

Wefer, G., Fischer, G., Fütterer, D., and Gersonde, R., 1988. Seasonal particle flux in the Brainfield Strait (Antarctica). Deep-Sea Res., Vol. 35, No. 6, p. 891-898.

Wetzel, A., 1983. Biogenic sedimentary structures in a modern upwelling region: Northwest African Continental Margin. In: Thiede, J. and Suess, E. (Eds.), Coastal Upweling, Its Sediment Record, Part B, Plenum Press, p. 123-144.

Williams, K.M., 1986. Recent Arctic marine diatom assemblages from bottom sediments in Baffin Bay and Davis Strait. Mar. Micropal., Vol. 10, p. 327-342.

Wilson, D.L., Smith, W.O., and Nelson, D.M., 1986. Phytoplankton Bloom Dynamics of the Western Ross Sea Ice Edge. I. Primary Productivity and Species-Specific Production. Deep-Sea Res., Vol. 33, p. 1375-1387.

Wolf, T.C.W. and Thiede, J., 1990. History of terrigenous sedimentation during the past 10 Ma in the North Atlantic (ODP-Legs 104, 105 and DSDP-Leg 81). Mar. Geol., in press.

Zafiriou, O.C., Gagosian, R.B., Peltzer, E.T., Alford, J.B., and Loder, T., 1985. Air-to-sea fluxes of lipids at Enewetak Atoll. Journ. Geophys. Res., Vol. 90, p. 2409-2423.

Zahn, R., Winn, K., and Sarnthein, M., 1986. Benthic foraminiferal δ^{13}C and accumulation rates of organic carbon: *Uvigerina peregrina* group and *Cibicidoides wuellerstorfi*. Paleoceanography, Vol. 1, p. 27-64.

12. Subject Index

Lecture Notes in Earth Sciences